The Inevitable Solar School

The Inevitable Solar School

Building the Sustainable Schools of the Future, Today

Mark Hanson

ROWMAN & LITTLEFIELD
Lanham • Boulder • New York • London

Published by Rowman & Littlefield
An imprint of The Rowman & Littlefield Publishing Group, Inc.
4501 Forbes Boulevard, Suite 200, Lanham, Maryland 20706
www.rowman.com

6 Tinworth Street, London SE11 5AL, United Kingdom

British Library Cataloguing in Publication Information Available

Library of Congress Cataloging-in-Publication Data Available

ISBN 978-1-4758-4419-1 (cloth)
ISBN 978-1-4758-4420-7 (pbk.)
ISBN 978-1-4758-4421-4 (electronic)

For
Reese, Katelyn, Sam, Charlie, and Christina
You will live in the world we leave you.

Contents

Preface   ix
Acknowledgments   xv

Introduction   1

**SECTION 1: CONTEMPLATING A NEW OR REMODELED
SCHOOL**   7

1   What Do You Expect in a New School Building?   9

2   A Report Card on Our School Building Stock   17

3   Where Do We Want to Go?   23

4   How Do We Span the Divide between Here and There?   31

5   The Inevitability of Solar Schools   43

SECTION 2: DEPARTURE   47

6   So, You Want a Sustainable Solar School   49

7   Encountering Fearful Buyers and Obstructionists   55

8   Solar PV Systems and Intelligent Schools   65

9   Technological and Financial Hurdles and Resources   77

10 Acquiring a Solar School   87

SECTION 3: FINDING THE MAIN CURRENT—CASE STUDIES   95

11 Northern High School   97

12 East Coast Elementary School 105

13 West Coast Middle School 111

14 Old Elementary School in the Sierra Nevada Foothills 117

15 Southern Elementary School 123

SECTION 4: WHAT LIES AHEAD **129**

16 Technological and Economic Forces 131

17 The River Ahead 137

Glossary 145

Index 149

About the Author 159

Preface

The places where we spend most of our time use almost 40 percent of the energy in our economy. These places are our residential, commercial, and for some of us, industrial buildings. As Americans, meaning citizens of the United States, are the largest per capita users of energy on the globe, our energy use in buildings represents a large resource use, a substantial operating cost, and a large share of our enormous carbon footprint.

A carbon footprint is a measure or indicator, and it's telling us that our human-made energy systems that support our living and economy are still heavily dependent on fossil fuels. It is the release of the stored carbon in our fossil fuels into the atmosphere in the form of heat-trapping CO_2 (carbon dioxide) that is the leading source of climate change.

Our understanding of the impact of CO_2 emissions on global warming emerges with the work of Svante Arrhenius at the end of the nineteenth century. He used the principles of physical chemistry to calculate the extent to which releases of atmospheric CO_2 increase the Earth's surface temperature. Charles Keeling subsequently was the first to conclude that CO_2 emissions from human activity were large enough to cause global warming. Since this early work, the scientific and popular understanding of global warming and its implications for climate change has evolved greatly.

A majority of us understand and are now beginning to dread what is coming in terms of the impacts of a massive and historically unprecedented rate of global warming. Global warming and associated climate change phenomenon including storm events are impacting natural ecosystems, water sources, agricultural systems, coastal and other areas subject to flooding, and the way we live.

The point is that climate change is a subject that matters greatly, and the present situation where the preponderance of our commercial buildings has large carbon footprints does not need to be. We are quickly coming to a point in time where we realize that we don't need that much energy from any source to operate our commercial buildings and the energy that is required can be provided by renewable energy.

In most cases, on-site solar PV (photovoltaic) systems that convert sunlight to electricity are the most advantageous source of renewable energy for a commercial building. Of great importance is the emerging reality that we can provide on-site solar energy on a cost competitive basis with fossil fuels. This assertion flies in the face of two common myths that we are having a difficult time disposing of. The first myth is that an energy-efficient building requires a large cost premium to build. The second myth is that on-site solar energy will come only at a large cost premium relative to power purchased from the electrical grid.

Parallel statements regarding adopting on-site solar PV or other forms of renewable energy can be made for residential and industrial buildings, as well as other non-building energy applications in transportation, agriculture, and industry. Those other prospects are fascinating, especially the transition to electric vehicles powered by renewable energy, but they are not the focus of this book.

This book focuses on schools and by extrapolation to other commercial buildings as this is the area where I can speak from my experience in the design and construction profession. I also draw on my experience in research, program development, and R&D funding. The experience comes from collaboration with school administrators, boards, and other building owners; occupants including students and teachers; and facility operators. It also comes from colleagues, both within the building design and construction communities as well as from colleagues in the energy research community.

My first degree was in economics. My area of interest is environmental economics. I was both fascinated and at times discouraged by how market forces, including externalities distorting the marketplace, influence human behavior. That interest in market forces and behavior is an important theme in this book.

I use economics to help interpret behavior in the marketplace for schools and other commercial buildings. The enormous decline in the costs of solar PV systems and energy efficiency in our sustainable buildings have enabled them to emerge as key features of a new, more-than-affordable, sustainable building approach as an alternative to the buildings we've gotten used to. Sustainable, solar-powered schools are better buildings across a spectrum of performance characteristics. As the school marketplace recognizes these advantages, public and private school boards, administrators, citizens, and school building designers and contractors will change their behavior.

This book is the story of school districts and other building owners, operators, and occupants, and their interactions with building professionals. The commercial building marketplace is an odd one. Many building owners such as school districts who need a new building or a remodeled building are inexperienced at buying or remodeling a building and at procuring on-site solar PV systems. Some have never procured a new building or a major renovation and a solar PV system. Others have done it before, perhaps ten or twenty years ago. Only a small proportion of school districts and private schools have considerable, recent experience. When there is unequal knowledge and experience in the marketplace between building and solar buyers and sellers, it creates conditions for less than optimal outcomes, or as economists would put it, inefficient outcomes.

There is all manner of difficulties in the building marketplace. If the people and organization wanting to build a school are the demand side of the market, who do they go to for the supply side? What do you ask for in terms of building size, configuration, and performance? What might be a realistic cost range for a school? How do you purchase the land on which the school is to be located, and what permits are required from various governmental entities? How are energy, water, IT, security, and other services provided?

Does on-site solar make sense and where should it be located? How much will it cost to operate the school in the coming years, and how long will the school last? These and many other pieces of information are what school buyers should know to determine if they can even afford to buy a new school or do a major remodeling. These are critical questions and building owners such as school districts are often lacking experience and are sometimes neophytes in the entire building acquisition process.

While the building buyer is often inexperienced in this complicated purchase, one might take for granted that those building buyers who routinely purchase buildings would be far more experienced and knowledgeable. This assumption, however, can be surprisingly misleading based on some of my experience even with owners who have procured dozens of similar buildings within a decade.

In contrast to the building buyer, the players on the supply side of the school building and solar marketplace are operating continuously in the market and are highly experienced. The supply side can be viewed as a supply chain with architects, engineers, general contractors, and skilled trades provided through specialized contractors such as mechanical, electrical, plumbing, fire protection, IT, and other specialties. It includes a myriad of material and product providers.

Using the parlance of economics, the market for commercial buildings is made up of often inexperienced and relatively uninformed buyers and highly experienced sellers and supply chains. This common asymmetry of knowledge in the marketplace has two implications. The first is the school or other commercial building buyers are often in a new, uncomfortable position. They are uncertain about what they are doing because they are not experienced in making this purchase. And they are facing building sellers and an entire supply chain that is highly experienced, at least in what they are used to selling. To be blunt, *highly experienced* is not to be confused with *high performance*, *highly refined*, *elegant simplicity*, or *sustainability* in buildings.

This asymmetry in the marketplace is not universal. There are some school districts and private schools with administrators, business managers, facilities staff, or board members with substantial experience and expertise in building procurement. Even in these cases, buying a new school or major remodel remains a daunting task.

Considering the implications of climate change, it is critically important that schools and other types of commercial buildings have low energy requirements for operation and have these requirements met with renewable energy, in most cases on-site solar PV systems. The slowness of innovation in the building market has helped to obscure recent trends in high performance buildings and solar PV systems that enable commercial buildings to operate with very low energy use and for much or all of that energy to be provided by solar energy.

The sudden emergence of cost competitive on-site solar PV since about 2010 is a critical turning point that has allowed financially competitive zero energy sustainable buildings such as schools. The term *zero energy* is interchangeable with *zero net energy* and *net zero energy*. It describes a building that produces as much energy as it uses over the course of a year and is connected to the electrical grid, at times buying power and at other times selling power.

The purpose of this book is to provide some rationale on why and some guidance on how school districts and private schools can procure a financially advantageous solar-powered sustainable school. My goal is to encourage school administrators, facility managers, business managers, board members, community members, teachers, students, and their building professionals to consider and then build a zero energy school. Another goal is to provide guidance to project team members in approaching and working through the long, complex building process. The book, however, is not a building manual.

To help explore the challenges, uncertainties, emotions, and group dynamics that are part of building a solar-powered school and other types of commercial buildings, I employ the metaphor of a river trip.

At the age of five, my parents bought a house on a small lake at the edge of town. The view from the kitchen window was of water and woods that eventually climbed up a large hill with a stone outcrop at the summit. Endless hours were spent in a wooden rowboat exploring every inch of shore of that pond, most of which was undeveloped land, including a hidden cove, and swimming at many locations on the lake. The change of seasons transformed our activities into skating and pond hockey.

The early introduction to the water and exploration evolved into a lifelong pursuit of water trips in remote places—the more remote the better. Canoe trips and occasional raft and dory trips on quiet and white water in Wisconsin, Minnesota, Ontario, Montana, Wyoming, Idaho, Colorado, Utah, and Arizona have become part of the fabric of my life. This travel has enabled me to see stunning beauty and experience life removed from the manufactured spaces created by humans.

River trips have lessons for individuals and groups. The ability to read a river and subsequently maneuver a vessel through all manner of obstacles and conditions as a group is a mesmerizing activity, thrilling and satisfying. Many of the sustainable building and/or solar projects that I have participated in or observed from a distance have some of the group dynamics of a river trip.

Thus, I use the metaphor of river running and wilderness lake travel to discuss the process of planning, designing, constructing, and operating a solar-powered, sustainable school. The metaphor also applies in describing the broad trends in the marketplace for building commercial buildings in general, including office buildings, medical clinics, assisted living facilities, and hotels. The current is carrying us to solar-powered zero energy sustainable buildings.

Some of the inevitable comparisons between river trips and building projects include considerations of whether you have a capable and compatible group on the trip. The challenge of reading the river and positioning the craft in suitable channels is somewhat akin to understanding the needs of the school district or private school, and creating a design to serve those needs. Both the river and the school project require team cohesion, especially with the owners. The outcomes of most river trips as well as a sustainable school project are in some respects foreordained in that a project will usually be built and the river will be run.

Every trip is different, and each project is different. Some hoped-for trips never get to the launching point or only partway down the river. A Wisconsin school district's recent RFP (request for proposal) for a set of large solar systems to be funded with third party investors in 2017 got upset in a rapid. The solar proposal did not proceed, at least so far, beyond the bidding stage.

The case studies that will be discussed in this book did move to successful completion, but sometimes only after being grounded in shallows or pinned in rapids. After some rescue and course correction, the projects moved on to successful completion.

This book is about making solar-powered, sustainable schools common. It is hoped that it will empower school districts, private schools, and other building owners of medical clinics and hospitals, retirement communities, religious communities, governments, and private firms with their offices and manufacturing facilities to ask for, or demand zero energy buildings with on-site solar energy.

Empowering buyers reduces fear and improves outcomes in buying, renovating, and upgrading schools and may even accelerate innovation. The sudden arrival of more than competitive on-site solar energy is propelling the trajectory of sustainability. It's helped to further define what we mean by sustainable solar buildings and makes them inevitable.

Acknowledgments

I have had the great fortune to learn from and work with wise and talented people. This book would not have been possible without them and the professional expertise and experiences they provided. Recognizing that I'm omitting many that should be acknowledged, I'll note some of these colleagues and friends.

Professor Wesley Foell of the University of Wisconsin and RMA has been a mentor since graduate school and reviewed early sections of the book. His ongoing insights and encouragement were useful, usable, and used. Niels Wolter of Madison Solar Consulting has been a co-worker on numerous solar projects. He provided guidance on solar PV systems technology, policy, market conditions, and performance. He also reviewed sections of the book. John Young is an innovator in third party investing in the Wisconsin solar market. As a collaborator in solar projects in Wisconsin and elsewhere, John has shared his knowledge in solar finance, tax treatment, and financial performance. Steve Carlson of CDH Engineering has collaborated as an energy modeler and commissioning agent. He reviewed sections of the book. Thomas Taylor of Vertegy Consultants has provided insights on the construction process especially as a collaborator in numerous LEED certified projects.

Hoffman Planning, Design & Construction Inc. provided the professional context and allowed me a wide latitude for innovation in sustainable commercial building projects, including schools. The planning, design, and construction management activities encompassed renewable energy beginning in 2004, PV systems beginning in 2006, and is now including batteries. Of the many colleagues at Hoffman who I've been privileged to work with, Paul Hoffman, Jody Andres, Tox Cox, Catherine Cruickshank, Pat Del Ponte, and Henry Hundt have been especially important in different ways as co-workers. Catherine provided a review of an early draft of the book.

School districts are an essential part of this story. Of note are District Administrator Dr. Mike Richie and Building and Grounds Director Dave Bohnen of the Northland Pines School District; District Superintendent Dr. Mary Bowen-Eggebraaten of the Hudson School District: and District Administrator Dr. Denise Wellnitz and Head of Maintenance Lee Black of the Darlington Community School District. A group of teachers at the Northland Pines High School led by Amy Justice, Robin Indermuehle, and Ann Perry provided leadership in making a sustainable school a living laboratory.

Prioress Mary David Walgenbach at Holy Wisdom Monastery has been a courageous leader for zero energy as part of the mission of care for the earth. She was the first client who asked for a zero energy building in 2008. I want to thank the publisher Rowman & Littlefield. Vice President Tom Koerner initially asked me to write this book based on an article I had written. Editorial staff including Emily Tuttle, Melissa McNitt, and Carlie Wall have been patient and steady professionals through this process. Finally, my wife Janis Hanson has been supportive and cheerful during my career and the writing of this book.

Introduction

This book is organized into four sections. The first section, Contemplating a New or Remodeled School, begins with a description of contemporary attitudes and expectations for schools and other commercial buildings. This is followed with an overview of the status of school buildings at the end of 2017, especially with respect to their comfort, energy use, cost, and carbon footprint.

It is from this baseline, that the prospects and rationale for widespread adoption of sustainable or green buildings with on-site solar PV systems are considered. Where are we in sustainable building practice—and particularly the adoption of solar PV? Where do we go next and why?

The commercial building market in most areas of the United States can provide a highly energy-efficient school, with a small portion of the energy use provided by solar PV, at a construction cost competitive with new schools with conventional but dated designs. That's a good start on a green school design. That's not sufficient, however, for the goal of a zero energy school where solar provides 100 percent of the energy on a net zero basis.

A zero energy solar school is a school with a solar PV system and is connected to the electrical grid. The electrical grid is the system of high-voltage transmission lines and distribution systems, including transformers and low-voltage lines, that connect power plants to electric power users. A zero energy school will export (i.e., sell electricity) when solar production is higher than the school can use at a given point in time, and it will import or buy electricity when solar power production is not enough to cover the electrical need. The school is zero energy if the sum total of export of power is the same as the import power over the course of a year.

A school becomes net positive if its power exports exceed power imports on an annual basis. Why do we want to build zero energy schools, and how do we do that over the next five to ten years and beyond? These questions will take us into the areas of intelligent buildings, intelligent grids, and electricity storage devices, notably batteries.

Section 1 concludes with consideration of how we approach the goal of a zero energy school and why it is an inevitable model for many, but not all schools.

In Section 2, the focus turns to the process and challenges of acquiring a zero energy sustainable school or acquiring a zero energy capable school when a school district is limited due to regulatory or other limitations—perhaps only temporarily—to providing only a portion of the school's annual energy from solar. The limitations are often not physical, such as having enough space to locate solar panels. Rather, they are often financial and regulatory, and these barriers are likely to eventually disappear in most states.

The initial perspective is that of a school district or private school wanting to procure a new building, remodeling project, or perhaps only a solar energy system for a zero energy capable school that already exists. In other words, this section considers the demand side of the school building market. It considers the often-daunting decision by school districts or private schools who are coming to the realization that they need or want a new school or major remodeling of an existing school. What function will the school provide? What features are desired and, critically, how does the school district and its leaders work through the critical task of deciding what they can afford or, better put, how to make it affordable?

Designers and builders sometimes joke about wishing to work with a client for whom money is not a high priority item. The underlying motivation for joking is not the added profit potential, but the freedom in design. Most designers have not met that client.

The deeper question is what should school districts ask for in their buildings? What should they be able to get in their school and at what cost? An important observation is that many school district administrators, business managers, board members, and community members don't even know what to ask for with respect to energy, sustainability, and construction cost. *And if you don't know what to ask for, how will you ever get what you really wanted in a new school or remodeling project?*

So, what should building and solar PV system buyers be looking for? How do they ask for it? And since they often don't know, how do they figure that out? What gets left out or left to chance? Section 2 provides perspective, information, and examples for school districts and other types of commercial building owners on how to procure a solar-powered sustainable school and help level the playing field in the school building marketplace.

Having first considered the buyer perspective, Section 2 then turns the focus to the seller side (i.e., the supply side of the school market). These are the building professionals such as planners, architects, engineering designers, and constructors who form the supply side of the commercial building and on-site solar PV market.

Building professionals can be selected to form an effective team with the building buyers. In successful projects, there is close collaboration on the team, or if you will, between the demand side and the supply side of the commercial building and solar markets. The collaboration logically includes some education by the building professionals for the school district or private schools and their building operators for when they occupy their new solar-powered school.

A river trip requires coordination and collaboration among the members who will be spending a lot of time together engaging in the activities of the trip. There are issues of skill and competency, organizational structure, trust, leadership, and agreement on many things ranging from the food and camp-sites to how to navigate each section of the river and the rapids encountered. Critical issues such as how to provide rescue need to be understood prior to when members of the team encounter upsets on the river.

"Who do you trust?" questions are a central and ongoing concern for a school district or private school as it procures a school. The relationship between the highly experienced building and solar-system professionals and the sometimes inexperienced, with respect to construction, school district or other building buyer is complex. The challenges and fears for a school district range from how to work through the myriad choices and tradeoffs involving project elements and costs, to potential conflicts of interest and the fear of just being taken.

The school building market is not unlike other markets such as health care, where highly experienced medical professionals engage their patient who's never had Lyme disease, a cancer, or fill in the blank for the malady. There are differences between the school building market and the health care market, including the prevalence of getting second opinions in medicine. Second opinions don't happen much in school projects. But as you are already imagining, trust matters. It takes time to build, must be maintained, and can be quickly shattered.

To help both buyers and sellers in the solar and school building markets, Section 2 explores the following topics:

- Common buyer fears
- Obstacles and obstructionists to zero energy sustainable schools, both real and perceived
- Building and solar technologies, components, and supply chains

• Financial issues, tools, and strategies
• Elements of successful projects

Section 3 transitions from a general presentation of the what, why, and how questions explored in Sections 1 and 2 to case studies of zero energy or near zero energy sustainable school projects. As we transition from the somewhat-efficient schools being built at the turn of the century to the high performance, zero energy capable buildings of today with substantial solar, on to the solar-powered zero energy sustainable schools we are rapidly and inevitably closing in on, the individual school projects will be the proving grounds for success.

The stories in Section 3 are about five zero energy or near zero energy schools that are examples of school projects that have gotten us to where we are today. These schools from different parts of the country in both urban and rural settings have much to teach us collectively, and much to teach school districts, private schools, and other commercial building buyers now thinking of what to ask for or demand in their next project. While these case studies and the book in general is focused on the United States, what we are learning is applicable internationally and we will learn from solar schools in other countries.

An important concept that is revealed in these case studies is the idea of a school as a living laboratory. A school need not be thought of as a stagnant set of technologies in a physical space with students and teachers who are not connected academically to the space they are using. Rather, schools can be viewed as dynamic test beds for utilizing evolving building and learning technologies to enhance building performance. The living laboratory approach adds resiliency to the school building and its use over time, with the goal of providing superior environmental conditions for learning. They become teaching labs for measuring and improving environmental quality, energy efficiency, and renewable energy.

School buildings and other commercial buildings are usually long-lived buildings that will go through a stream of changes in their lifetimes that sometimes extend into the hundreds of years. Emerging technologies can be incorporated over time to improve building performance and, with solar PV, increase energy production. The widespread adoption of LED lighting, smart building controls, and solar energy has occurred over the last decade, while batteries are just beginning to arrive. School districts with their teachers and students can use their entire building as laboratory and learning test beds.

One of the case studies is a sustainable high school completed in 2006 with minimal on-site solar because the cost of solar was exorbitant at the time. The project team counseled the school district to wait for solar PV system prices to fall. They did, and a large solar system was added eleven years later on a cash flow positive basis. For this school district, however, this may only be another step as it anticipates additional solar, batteries for energy storage, and building intelligence that will eventually transform the school to zero energy.

Another case study is an old, small elementary school that has been remodeled to zero energy. Two other case study schools were conceived of and constructed to be zero energy schools from the start. The remaining case study school was designed with at least ultra-low-energy use in mind if not zero energy and constructed that way. While it has considerable on-site solar energy, it is anticipating future batteries and needs some additional solar energy to reach zero energy. The case studies represent a cross section of geographic and climate conditions.

A theme throughout the case studies are the financial issues and how they were managed by owners with real-world financial constraints. A stubborn myth is that if you want a green building, reach for your wallet, as you'll need it. The case studies are examples where solar-powered, sustainable schools were acquired at cost points comparable to or less than conventional costs. This includes some solar systems financed with or owned by third party investors and are cash-flow positive with no first cost to the school.

Section 4 looks ahead. It considers the pace of technological change in solar and green buildings and what will be emerging in the next five to ten years, and beyond. Technological change is interwoven with financial factors, such as tax policy, as well as environmental factors, such as climate change consequences and the resulting policy responses. Technological change is being driven by forces on a global scale, with Germany and China particularly important players in the solar market. In the United States the 30 percent federal investment tax credit, accelerated depreciation, and low interest rates have been recent drivers of third-party financing of on-site solar for not-for-profit commercial buildings such as schools.

For-profit entities have taken advantage of the tax treatment as well, noting that according to information compiled by the Solar Energy Industry Association, many firms have invested heavily in solar energy either with their own funds, or through third-party providers. Solar installations in MW (megawatt or 1,000 kilowatt) as of 2016 for some leading firms adopting solar include Target (147 MW), Walmart (145 MW), Prologis (108 MW), Apple (94 MW), Costco (51 MW), Kohl's (50 MW), and IKEA (44 MW).[1]

Energy storage and intelligent buildings and grids are the emerging technologies that will support the emerging zero energy solar schools that will become commonplace—and eventually dominate the market. The commercial building stock is large and slow to turn over. Thus, the full implications of zero energy solar buildings will take a long time to play out.

Section 4 closes with consideration of the potential for solar-powered sustainable schools and other commercial buildings in the overall climate change picture. The hope is that the movement to solar zero energy schools will still matter, and that the stories shared in this book provide inspiration and guidance for many zero energy sustainable schools and other commercial buildings. The challenges of working through the creation of new zero energy schools and the transformation of existing schools to zero energy may be thought of as river trips and enjoyed immensely.

NOTE

1. Solar Energy Industries Association, 2017.

Section 1

CONTEMPLATING A NEW OR REMODELED SCHOOL

Chapter 1

What Do You Expect in a
New School Building?

The launch of a new school project is an exciting time of looking forward, with high expectations and perhaps even higher trepidation. There is lots of planning to be done, not the least of which is gaining community support to fund the project. The building should serve the academic mission and other community goals, provide prestige to the school district and broader community, provide an inspiring and flexible built environment in which to learn, and operate at affordable levels in terms of the maintenance effort and the electricity, natural gas, water, and information technology (IT) costs.

The planning challenges are similar whether the school project is public or private. The definitions of community will be different as will the sources of funding, but the underlying goals of how the school should function and serve its users will be similar.

The average age of public schools in the United States is 44 years, with an average of 12 years since major renovation.[1] This means that many years may have transpired since small school districts and even some large school districts have gone through the building process. The school administrators and school boards often don't have recent experience in procuring a remodel or new facility. In other words, school districts tend to be inexperienced buyers.

Large and especially rapidly growing school districts will generally have more recent experience. In some of these cases however, they may not be at the cutting edge of school design because of the tendency of school districts to procure a new school and then replicate the same building type and technologies without making the effort to learn from the responses of students, teachers, facilities staff, and other building users to their recent school projects. Given the numerous demands on school districts, they may pay scant attention to building performance such as energy use and cost, water use and cost, indoor air quality, internal noise, and maintenance experience.

Barring a disastrous recent purchase experience, these more frequent buyers often return to the supply chain they purchased recent school buildings from, because they are now familiar with those providers, and they proceed to buy the same or similar buildings over and over. The supply chain providers include the following:

- Planners
- Architects
- Landscape architects and site designers
- Civil engineers
- Structural engineers
- MEP (mechanical, electrical, and plumbing) designers
- Construction managers or general contractors
- Various contractors ranging from structural to mechanical, electrical, and IT
- Numerous suppliers of materials and equipment

Thus, even in the case of repeatedly procuring a school, the buyer may have a limited knowledge of the latest or emerging opportunities for their buildings.

The commercial building planning, design, and construction market has a specialized vocabulary. Introducing some of these terms at the outset will save time and confusion. Additional terms will be introduced during the course of the book and a glossary that includes a list of acronyms is provided.

The most confusing systems and terminology related to schools are their mechanical systems. What are the mechanical systems? These are the systems that control temperature, humidity, outside air volume, and overall air circulation within the school. They consist of boilers, chillers, fans, pumps, ductwork, and so on, and their associated control systems that appear in many configurations. Even though some school districts or even designers will occasionally replicate a design, most schools in the United States are unique, customized buildings.

Mechanical systems provide a source of heat for virtually all schools in the United States, and a source of cooling for many schools. Cooling, which is also commonly called air-conditioning is usually provided through compressors that run on electricity. Heating has historically been provided by natural gas boilers, although electric resistance heating is found in some areas and heat pumps are gaining in popularity.

A heat pump is a device that transfers energy from a heat source to a heat sink. In a school, it uses a refrigerant to transfer heat from, say, outdoor air or from the earth and releases the heat into a space to be warmed, such as a classroom. Its operation is similar to a refrigerator where heat is removed from within the refrigerator and released into the kitchen. Heat pump systems with their compressors are more expensive than simple electric resistance heating, such as baseboard heaters, but are up to three to four times more efficient in the use of electricity to provide heat.

Heat pumps may also be configured to be reversible to provide cooling for indoor spaces while transferring the heat outdoors or into the earth. When the heat exchange is with the earth, heat pump systems are called geothermal. Closed-loop well systems are commonly used to transfer heat to and from the earth.

Building codes set the requirements for conditions within schools and minimum performance requirements for mechanical equipment and many other aspects of school buildings. Building codes are determined at the state level, but much effort has gone into developing national and international standards that are then adopted, sometimes with modifications, by states.

Mechanical systems need to respond to a wide range of changing meteorological conditions, as well as internal conditions, such as the number of people in a room, and other loads in the room, such as lighting and computers. The controls of the mechanical equipment have become increasingly sophisticated with the application of computers. The systems are commonly referred to as buildings automation systems (BAS), building control systems (BCS), or heating, ventilation, and air conditioning (HVAC) controls.

The bottom line in designing, building, and operating mechanical and lighting systems is to provide a high performance (i.e., a highly energy-efficient) school that provides a comfortable and highly controllable indoor environment conducive to learning. The American Society of Heating, Refrigerating, and Air-Conditioning Engineers (ASHRAE) is the professional association at the forefront of the science and engineering of mechanical equipment to provide temperature, humidity, air quality, noise levels, and light levels in an efficient and safe manner.

The process of purchasing a new school or the remodel of an existing school has multiple layers of complexity that are intertwined with the financial reality that the cost ranges from approximately $20 to $30 million for a moderate to large new elementary school of 100,000 square feet. For a moderately sized high school of 200,000 square feet, the cost ranges from $45 to $75 million, excluding land costs. There are large cost disparities across the country, with some low-cost areas in the South and Midwest falling below the ranges noted and high-cost areas along the East and West Coasts and some large urban areas in various parts of the country coming well above these ranges. Other complexities to be worked through include the nature and flexibility of the learning spaces and the sustainability desired in the design, construction, and operation.

The USGBC (U.S. Green Building Council) LEED® (Leadership in Energy and Environmental Design) rating system introduced in the early

2000s, along with other green rating systems such as the Green Building Initiative's Green Globes© and the International Living Future Institute's Living Building Challenge, have had the anticipated transformative impact on the building industry in terms of making design, construction, and operation increasingly green or sustainable. On the one hand the use of green ratings systems has added another layer of consideration in the school buying process. On the other hand, it has helped to define what is meant by sustainable and green and provided a growing wealth of sustainable services and products to the school design and construction process.

Perhaps the greatest complexity in the school buying process comes in determining and negotiating the project price. The buyers who have not purchased a new school in the last ten years have a great disadvantage in the marketplace as they don't know the costs for new school or school remodeling projects, especially in the face of a plethora of design choices, materials, components, and contracting options. Further complicating this challenge for the buyer is ascertaining and choosing the greenness of the facility and sorting out underlying greenness from the greenwashing claims. Third-party certification programs such as LEED are helpful in identifying and managing green choices.

Important aspects of sustainability are not only the ongoing energy requirements and costs, but the sourcing of that energy. On-site solar energy is trending as it not only can provide energy for a school at lower costs, but it also changes the energy sourcing for electricity away from power plants that in many areas of the country heavily rely on fossil fuels. On-site solar energy combined with energy efficiency also help to reduce the energy infrastructure required to deliver energy to the school from central power plants.

The fossil-based energy supply chain infrastructure starts at mines and oil and natural gas fields with fuel extraction, processing, and transportation to central power plants. Power is then moved via the transmission and distribution system to customers, including schools. While the central power grid is rapidly adding renewables, especially wind and more recently solar, it remains heavily dependent on fossil fuels such as natural gas and coal in many parts of the country.

First appearances suggest that on-site renewable energy will be a heavy lift in that renewable energy, such as solar PV systems, is assumed to require further up-front investment beyond the school building itself. Third-party investment for on-site solar, however, can provide solar electricity in a manner that is comparable for a school operating budget to the monthly payments to the electric and natural gas utilities. Third-party investment occurs when outside investors own and install the solar PV system at a school. The school typically pays for the solar power monthly as it would for utility power. Some third-party investment arrangements include eventual ownership transfer of the solar system.

By 2015, the use of third-party investment for on-site solar had become the most common pathway for commercial-scale solar in the United States. Of commercial-scale solar installed in 2015, 65 percent was third party owned.[2] The remaining thirty-five percent was customer owned.

There are various approaches for third-party investment in solar PV systems for schools. Some of these options, which are discussed in later chapters, are not allowed in some states. The challenge for school districts and private schools is getting acquainted with what third-party investment options are available in a given situation and locations. And like many other aspects of a school project, there is much to be figured out in terms of advantages and disadvantages.

This is an example where the school buyer may be inexperienced in the face of a major purchase with many complexities. The buyer is in a marketplace where the sellers and their supply chains are designing and constructing solar PV systems continually. One way of describing this situation is that knowledge and experience in the school building market is asymmetric. In other words, there is a less-experienced buyer facing a highly experienced seller.

The asymmetry in the school building marketplace as well as many other commercial building markets is further cause for trepidation and uncertainty in building buyers. Buyers' fear and angst begins at an elevated level as they start the procurement process. The process typically starts with a competitive bidding process for planning services. If the project stays on track, this leads to bidding for design and sometimes for design and construction management services in combination. Trying to decide what firm to believe and trust as they make their pitches is no easy task and may feel like reading tea leaves.

As the planning and design process ensues over many months, queasiness or anxiety often grows, resulting from the many decisions that need to be made. Should you use a VAV (variable air volume) mechanical system served by central boilers and chillers, or would a distributed geothermal heat pump system be a better choice? Should the project include on-site solar energy, and if so, how much and should it be third-party financed? Will these systems be designed optimally and sized correctly? Will the building operate as intended? Will the buildings be reasonable to maintain?

There are a variety of common and, on the face of it, surprisingly frequent problems found in new school and other commercial building projects including these:

• Oversized mechanical systems
• Excessive lighting levels
• High contrast or glare from windows

- Mechanical systems not operating at their potential when the school is turned over to the school district, often due to the lack of commissioning or effective commissioning

  In the face of these reoccurring issues that the buyer typically assumes the building supplier would be able to consistently manage in the planning, design, and construction process, it's no wonder that the buyer may feel overwhelmed by the complexity. As the future building owners make their way through the necessary decisions, the anxiety level finally seems to decrease—at least in many projects—as the building approaches construction completion. The school district and its administrators can finally see the school taking tangible form.

This is not to say that a significant portion of school building projects come out poorly. It should come as no surprise that there is a wide range of project outcomes ranging from those that are high performing, sustainable buildings to poorly performing and not very sustainable buildings. The permutations and combinations are many. One example is an otherwise quite sustainable school, but with excessive lighting and associated lighting energy use. Another example might be a sustainable school design, but with excessively sized mechanical systems that impinge on effective operation.

If the building and solar design and construction processes have gone well across most or all aspects of the project, anxiety fades and the mood transitions to celebration as the students, teachers, and visitors first enter their new facility. If the building has been well designed and is operating as intended, neither of which can be taken for granted, fear gives way to what can be termed comfort and joy. If the design was poorly designed and/or executed for the buyers' needs, which then results in operational problems in the early months that cannot be resolved in a timely manner, fear moves on to further mistrust, sour grapes, and finger pointing as to who is at fault. Was the design deficient or was the construction deficient?

The most startling outcome of the complexity in the process of buying a new school and the market asymmetry is that school buyers—that is, the school administrators, business managers, facilities staff, and boards—sometimes don't know what they want. When they do know what they want, they're not sure of how to ask for it when they go to buy a new school.

This blunt statement is specifically true regarding a set of important performance and cost criteria including the following:

- Energy performance
- Daylighting and lighting performance
- Acoustical performance, including background mechanical noise
- Indoor air quality

- On-site renewable energy
- Overall project cost

So, what should school districts administrators and boards ask for in a new school or major remodeling project?

The answer to that fundamental question, at its simplest, would be something that would include the following expectations, or perhaps better put, requirements:

- Provides flexible learning spaces to meet the academic mission and community goals
- Energy performance measured in the top 25 percent of schools
- Classrooms that provide daylight and views, supplemented by electric lighting as needed
- Exceptional indoor air quality as measured by the six criteria used by LEED (formaldehyde, particulates, ozone, total volatile organic compounds, target chemicals, and carbon monoxide)
- Minimum acoustic performance per LEED for background noise, reverberation, sound transmission, and exterior noise
- Cost per square foot that is at or below the conventional cost average in the region
- On-site solar PV energy that provides a substantial portion or all the energy required
- An intelligent building in terms of the ability to sense and respond to building usage, solar production, and energy price signals

The reader might wonder at this point whether this ask is reasonable or even feasible. This is especially true given the common myth that sustainable or green buildings will cost more than conventional buildings.

There are many examples where sustainable buildings have been built at first cost levels that are competitive with first cost in comparable conventional buildings with higher energy use and no solar energy. There are examples of schools where substantial or even zero energy amounts of on-site solar energy have been included that are more than cost competitive with conventional schools with no solar energy.

The barriers to zero energy solar schools have been removed in some states. In other states, moving to 100 percent on-site solar PV involves working around electric utility regulatory barriers. Emerging battery storage and intelligent grids will make zero energy schools that are first cost competitive even easier, including in states with regulatory barriers.

As will be discussed, there have been some studies comparing LEED certified buildings to non-certified buildings of similar types. One finding of those studies was that there is a large range of costs for the same building type after accounting for year of construction and geographic location within groups of both the certified and non-certified buildings.[3] A second finding was that there is no average cost differential between the certified and non-certified buildings. If LEED certification is an indicator of sustainable or green buildings and non-certification is an indicator of a more conventional buildings, experience shows that there are no added costs in going green. This is a remarkable finding.

If these statements on cost-competitive, sustainable, solar-powered schools are true, why are school administrators and districts not asking for a solar-powered sustainable school when they engage in the marketplace? This is perhaps a matter of lack of knowledge and experience to know what to ask for and long-term practices in school procurement.

The traditional pattern in schools is to work through planning and design processes. It's generally an iterative approach where the academic needs are initially determined, initial solutions provided by designers, and potential construction costs estimated. Final cost is determined later in the construction bidding process and in subsequent negotiations, including when the building costs must be reconciled with referendum budget limits.

Which of the asks listed above are included in a given design is a matter of what the design team has included as a matter of their usual practice. Practice is all over the map, and there are generally no incentives to specifically meet these asks, unless the school district asks for them.

If school districts and private schools or any other type of building buyer wants something in the marketplace, they are far more likely to get it if they ask for it. The flip side of this statement is that you aren't very likely to get this outcome if you don't ask—and insist if necessary.

NOTES

1. School Facilities at a Glance, *Education Week*, Vol. 37, Issue 14, November 29, 2017.

2. Greentech Media Research, *U.S. Commercial Scale Landscape 2016–2020*, May 2016.

3. Peter Morris and Lisa Fay Matthiessen, *Cost of Green Revisited: Reexamining the Feasibility and Cost Impact of Sustainable Design in the Light of Increased Market Adoption*, San Francisco: Davis Landon, July 2007.

Chapter 2

A Report Card on Our School Building Stock

The existing school building inventory in the United States mirrors the report cards of the students these schools serve. The performance of these schools as environments in which to learn is all over the map with grades ranging from A to F. Some important sustainability criteria on which grades might be given include the following:

- Thermal performance
- Visual performance, including lighting and views
- Acoustical performance
- Indoor air quality
- Energy use and cost
- Water use and cost
- Maintenance cost levels
- Safety

There are many other criteria to consider, including IT capability, aesthetics, and types and adequacy of learning spaces ranging from media centers to music rooms, classrooms, and athletic facilities.

The structures themselves range from old buildings well more than 100 years old to new facilities, some of which are state of the art and some quite plain vanilla. Some new schools are sources of inspiration, others are mediocre, and some dismal. The mix of schools includes multistory buildings, single-story buildings, single-story classrooms that exit directly to the outside, and portable classrooms.

There were approximately 98,200 public schools in 2017, including 6,700 charter schools.[1] There were an additional 34,600 private schools. The total square footage for all schools is more than 8 billion square feet. The annual operations and maintenance spending for the public schools averaged $46 billion over the 20-year period from 1994 to 2013.[2]

Data provided from the U.S. Environmental Protection Agency's (EPA) Energy Star Portfolio Manager for energy use in K–12 schools covering 5.8 billion square feet of schools shows a median school size of almost 75,000 square feet.[3]

The Wisconsin Focus on Energy Program sponsored an extensive, voluntary survey of public schools in 2006.[4] More than 60 percent of Wisconsin's K–12 public schools participated. These 1,293 schools served 546,000 students, had 109 million square feet of building area, a mean school size of 84,000 square feet, and incurred energy costs exceeding $98 million per year. Of these schools 46 percent had partial or full air-conditioning. Energy use per square foot commonly varied by a three to one ratio in these schools, and by five to one in more extreme cases. When the energy use per square foot is plotted relative to the age of the school, there is no evident pattern. Newer schools were no more efficient with respect to energy use than schools 50 or 100 years old.

The question you may be asking at this point is Why don't the schools using higher levels of energy manage their usage down toward those using a half or even a third of the amount of energy? The answer is that some are, but it takes sustained effort in the face of the many other responsibilities of building operators and maintenance staff. The first responsibility for maintenance personnel is keeping everything running and responding to issues and complaints as they emerge.

Another answer is that energy use is low in some schools because of low performance, including factors such as inadequate provision of outside air to rooms and no air-conditioning in spaces in conditions where it would be beneficial. There is a difference between energy efficiency and poorly performing buildings that have low energy-use levels because of how they are operating.

The economic context is also important in understanding why energy use may not be a top priority in some schools. While every dollar of energy saved is a win for the school, energy costs represent on the order of only 2–3 percent of the school budget. Labor costs for teachers, administrators, support services including food and transportation constitute more than 80 percent of a school's budget.[5] It makes all the sense in the world for a school to make instructional issues and students the highest priority. Having said that, however, providing the best learning environments and lowering energy costs cannot be ignored as the former is an asset in student and staff well-being and performance and reducing energy costs takes at least some pressure off the budget.

The Energy Star Portfolio Manager provides a metric that is useful in evaluating energy use in schools. Energy units can become a complicated aspect of considering school performance and sustainability. A brief look at a few industry standard energy measurements is useful to set the stage for considering where schools are today and where they might want to go in the future.

A basic measure of school energy use is the Energy Utilization Intensity (EUI), which is measured in two ways. Source EUI is measured in kBtu/ft^2 per year. That is thousands of British thermal units per square foot of school building floor area per year. *Source* means the energy content of the fuel at the point of extraction, for example at a coal mine where the coal is mined for burning at a power plant that generates electricity. The other measure is Site EUI. *Site* refers to the energy content also in kBtu/ft^2 per year of the energy as it is delivered to the school property.

The biggest difference between Source EUI and Site EUI occurs when measuring the energy content for electricity provided to a school. Site energy measures electricity as the heat content of one kilowatt-hour (kWh), which is 3413 Btu per kWh. Because of energy losses from taking a raw fuel such as coal and converting it to electricity, the energy content of the source fuel is much higher, more than 10,000 Btu per kWh delivered to the school. The use of the Btu is a matter of our history as a British colony and our unwillingness to replace British or Imperial units with the metric system.

The two dominant forms of energy delivered to schools are electricity and natural gas. Fuel oil and propane gas are used for heating in areas where natural gas pipelines are not present. The electricity provided includes large efficiency losses at power plants when fossil, biomass fuels, or nuclear fission are used for the generation of power and much smaller losses in the transmission and distribution of electricity from power plants to power users. The sources of energy for generating electricity include coal, natural gas, nuclear fission, fuel oil, and renewable energy from wind, solar, hydro, and biomass or biogas. In the case of a coal-fired, other fossil, and nuclear power plants, roughly three units of energy are required for every unit of energy leaving the plant in the form of electricity.

The median Source EUI for the 55,000 schools in the Portfolio Manager is 114 kBtu/ft^2 per year with the 5th percentile (lowest 5 percent of energy users) at a Source EUI of 56 kBtu/ft^2 per year and the 95th percentile (highest 5 percent of energy users) at a Source EUI of 208.[6] This information is derived from 55,000 schools accounting for 5.8 billion square feet.

Virtually all schools in the United States use a heating source, most commonly natural gas. Some schools at more isolated locations may use propane or fuel oil for heating. Some schools use electricity only, including those providing heating and cooling with heat pumps that may be exchanging heat with the earth (geothermal) or with the air (air sourced). In addition to a heating source, schools and other buildings use electricity for lighting, ventilation, cooling, IT equipment, food preparation, and so on.

The adoption of solar energy in U.S. schools appears to be following what is called a *diffusion model*.[7] Solar adoption in schools has emerged from a small number of innovators into a larger number of early adopters and then on to an even larger number of schools in an early majority phase. A Brighter Future: A Study of Solar in U.S. Schools by the Solar Foundation, Generation 180, and the Solar Energy Industries Association (SEIA) found that 4.4 percent of schools serving 7.3 percent of students had on-site solar systems.[8] The surge in school solar adoption can be seen in solar offerings and training events such as the Zero Net Energy School Retrofit Trainings sponsored by the California Public Utilities and organized by the New Buildings Institute.

The 75 percent drop in the cost of installed solar PV systems since 2010 is accelerating solar adoption in schools. The dramatic decline in costs and the expansion of more favorable utility regulation regarding feeding excess power onto the electric grid is moving solar installations from smaller demonstration projects primarily for educational purposes to larger projects for operational cost savings and educational purposes. Educational aspects of solar energy are often linked to science, technology, engineering, and mathematics (STEM) but also include business, finance, environmental studies, political science, and fine arts.

The A Brighter Future study found that the average solar PV (photovoltaic) system size being installed at schools in 2010 was about 100 kW. The industry convention is to use kW-dc or direct current ratings on PV panels and PV system sizes. By 2014, the average size had increased to about 200 kW, and in 2017 the average size was near 300 kW. The funding had shifted from ad hoc fund-raising through incentive grants, fund-raising, and school budgets to a reported 90 percent use of third-party funding using PPAs (power purchase agreements) and other third-party investment (TPI) arrangements.

In part, the data reflect some school districts moving very decisively into large-scale solar using PPAs. The Kern High School District in California, for instance, has used PPAs to install 24,500 kW of solar at 27 sites in the school district.[9] The Solar Foundation reports that this is the largest contracted commitment to solar by a school district in the United States. The financial savings to the school district is reported at $80 million over 25 years. Considering that perhaps 500 kW of solar PV would be sufficient to provide for all the energy needs at the median school size in the Energy Star Portfolio Manager if the school were energy efficient, the Kern High School District is making solar a major part of their energy supply.

Energy efficiency and on-site renewable energy perhaps sufficient to achieve zero energy are essential elements of a sustainable school. There are, however, many other building characteristics to consider in creating sustainable schools—such as indoor air quality, acoustics, water use, daylighting and lighting, and views.

As in energy use, there is a wide range of conditions across these building characteristics from highly performing schools, including some that are LEED certified and some that are not, to schools that perform poorly. What's remarkable is that while building codes have been gradually improving over the last decades, there is very limited ongoing monitoring of existing schools to see if buildings that have been built to code at the time of construction operate consistent with the code over time. The USGBC, to its credit, has provided emphasis on monitoring in new projects and for existing buildings through its LEED Building Operation and Maintenance certification. These and other offerings in monitoring performance are gradually expanding.

In communities and states where education budgets are under revenue squeezes, it's challenging, to say the least, to get the resources in building operations and maintenance to hold the line on building performance, not to mention get the resources to make more costly repairs and modernization over time. The disparity in schools that are high performing and sustainable relative to schools that are being starved for resources and/or missing out on maintenance for whatever reason is evident to even the casual observer that gets the opportunity to visit a diversity of schools. It's easy to tell when the air is stale and CO_2 levels are above say 1000 ppm, there is way too much glare coming from the windows, the mechanical systems are noisy, the temperature is not well controlled, and the room design and materials do not absorb sufficient noise. It's easy to spot the older plumbing equipment with higher water-usage rates along with the leaks.

The evaluation of our existing schools and what we want to do has some analogy to river trips with respect to evaluating rivers. There are rivers that are appealing in their water quality, the quality of the shorelines and views, their remoteness, the adequateness of stream flows, and the desirability of the campsites. For those seeking white water, the rating level of the rapids becomes a highly important criterion for both the challenge and enjoyment—and for safety.

Rivers that fail the grade on some of these criteria are the ones that don't get run or get run less often. The difference between evaluating a river for river travel and schools for their performance and sustainability, however, is that we perhaps can walk away from the rivers that rate poorly in our criteria. In the case of schools, communities cannot walk away. Any school with failing grades should be improved or, if need be, replaced with a sustainable school.

What grade should we give to our school buildings? That's perhaps not the most useful question. The more useful questions relate to some of our pioneering examples of schools that are inspirational to attend and have shown how to achieve low energy use (i.e., low Source EUIs) and do so on conventional budgets. It's also useful to look at the early adopters of solar that are providing the energy required while achieving financial savings.

The more useful question becomes What grade should we be aspiring to and expecting in our future schools and school remodels? And what are the financial implications as we set those expectations?

NOTES

1. National Center for Education Statistics.

2. School Facilities at a Glance, *Education Week*, Vol. 37, Issue 14, November 29, 2017.

3. "Data Trends: Energy Use in K–12 Schools," Energy Star, January 2015.

4. Wisconsin Focus on Energy, Wisconsin Public Schools Benchmarking Project Case Study, 2007.

5. "School Budgets 101," www.aasa.org: American Association of School Administrators.

6. "Data Trends: Energy Use in K–12 Schools."

7. Malcolm Gladwell, *The Tipping Point: How Little Things Can Make a Big Difference*, Boston: Little, Brown and Company, 2002. Chapter 6 provides a popular treatment of this subject.

8. The Solar Foundation, Generation 180, and SEIA. *Brighter Future: A Study of Solar in U.S. Schools*. Second Edition, November 2017.

9. www.sagerenew.com. Selected case studies: Kern High School District Solar PV PPA.

Chapter 3

Where Do We Want to Go?

The conventional answer to this question might seem to hinge on an economic calculation. The economic calculation exists within the broad context of how to build or remodel a school to best serve the educational mission of the community and its students. If schools are overcrowded or in such a deteriorated condition that replacement is the reasonable course, a new school can be considered. The common perception is that if a school district is not satisfied with the results on its report card for an existing school and wants to improve the learning environment by providing a high-performing, sustainable school, it will need to decide what improvement it wants and be willing to pay the price.

Although deciding where to go on river trips is of much less consequence than deciding on educational missions and how to best provide schools to support those missions, a group of friends deciding on a river trip face the fundamental questions of where does the group want to go? There is a whole world of possibilities as to the types of rivers, lakes, or oceans to travel and explore. The decision for any group of river travelers will be unique to the group just as the decisions on the school solution will be unique to a school district and its needs, its community, and its context.

River trip options get narrowed by circumstances and experience. How much time do the group members have to be away? What kinds of boats do they have and want to use? What preferences does the group have for different types of water? Are they capable of running the rivers or traversing the lakes being considered? What types of environments do they want to explore? Finally, does the group have the financial resources to do the trip?

How much we should be willing to pay, and hence where we want to go—or perhaps where we can at least get to—are heavily driven by the economic concept of marginal costs and benefits. The economic calculation in

23

the case of schools considering energy use, as an example, is that there needs to be benefits, such as lower energy operating costs, to justify the added cost of a better-performing building relative to a lesser building. A similar rationale can be applied to many other benefits such as improved air quality, daylighting and lighting, noise management, water-use reduction, types of learning spaces, and other design features that all contribute to the learning environment.

These economic calculations are taking place within the context of the ongoing struggle to gain taxpayer approval to obtain tax dollars for public schools, or donors and perhaps increased tuition for private schools.

School districts are long-term organizations with long-lived facilities. This is the case with private schools as well. School districts tend to accept longer payback periods of ten years or more for improvements in existing schools, as well as in making choices for new schools. The fact that school districts are long-term owners and occupants changes the economic analysis to a much longer perspective than some other types of commercial buildings that may be occupied by entities other than the owners and may also be flipped if justified by short-term profits.

Segments of the commercial building market are plagued by a frequent problem that economists call *split incentives* where, for example, the original building owner is a developer, whereas the occupants pay the utility bills. The developer or subsequent owner may want the lowest first cost in building an apartment building and therefore diminishes the design knowing that renters or subsequent buyers are going to be paying the utility bills.

Fortunately, school districts usually don't face this dilemma. They can consider energy efficiency approaches and features that can be included in the school design at a cost, and that this cost is usually acceptable if the added cost can be covered by savings within a reasonable amount of time.

There are many factors that are considered in evaluating what constitutes a highly conducive and sustainable learning environment. Of the many factors that impinge on both operating cost and environmental consequences, energy use is a priority concern in school facilities operations and has been so historically. While energy costs are small compared to the size of labor costs in a school budget, it is sizable within the context of ongoing facilities operating budgets and a reminder comes every month in the form of a utility bill.

An advantage of managing energy as a sustainability criterion compared to some of the other sustainable criteria is the ease of measurement. A kWh of energy saved can be easily measured and assigned a monetary value, even though that ignores external costs such as carbon emissions. Attempting to measure and then valuing an improved learning environment, such as improved air quality, is a much tougher proposition.

Where we want to go with our school buildings and what might be achievable regarding the energy and associated cost in schools and other buildings is informed by multiple considerations including the physical condition of the school and the utility rates of the serving electric and natural gas utilities. Basic considerations include the condition of the shell of the school—in other words, the walls, windows, roof, and foundation. After the shell condition, the condition of the MEP (mechanical, electrical, and plumbing) systems that provide heating, cooling, ventilation, lighting, and indoor air quality are of paramount concern.

In the case of an existing school being considered for remodeling, the school's Source EUI will reflect both the condition of the school and how it is being operated. If an existing school has a Source EUI approaching the worst 5th percentile (208 kBtu/ft^2 per year), there is a lot of improvement potential and perhaps some low-hanging fruit in the form of lower-cost changes to reduce that number closer to the school median of 114 kBtu/ft^2 per year. If an existing school is already near the median energy-use level or even lower, it will tend to get more difficult to achieve energy-use reductions.

Regardless of the starting point, improvements and the energy cost savings that come with it will be weighed against some payback criteria. These, at least, are the conventional marginal cost and marginal benefit criteria resulting from conventional economic analysis.

There are seemingly counterexamples of new schools, however, that turn the marginal cost and benefit calculation on its head. Notably, there are high performance sustainable schools that provide outstanding learning environments, low energy requirements and costs to operate the schools, and were constructed at costs that are the same as or lower than the average, or shall we say conventional, schools. In other words, these sustainable schools seem to have gotten around the incremental cost for higher performance. The presumption that higher performance must cost more in a new or remodeled school does not seem to hold.

An example of a high performance school that was delivered at a construction cost much lower than conventional schools based on regional cost averages is River Crest Elementary School in Hudson, Wisconsin. It was built in 2008 near the twin cities of Minneapolis and St. Paul. River Crest's sustainability credentials include the recognition that it was the second certified LEED Gold elementary school in the United States.

The Source EUI in its first year of operation was 110 kBtu/ft^2 per year compared to the current national median described in chapter 2 of 114. At first impression, the River Crest Source EUI seems rather ordinary, close to the national median. What that does not consider, however, is the cold winter and hot summer climate, its use for summer school, and other factors. The Hudson School District uses River Crest for summer school programs in the district because of the amenities it offers and the air-conditioning. This explains Energy Star Rating Score well above 80.

River Crest's Source EUI falls within the range of Source EUI's for schools in the New Buildings Institute's Getting to Zero list of verified and emerging projects.[1] The range of Source EUI's for sizable schools on the list ranges from a low of 48.8 to a high of 160.7. Thus, River Crest is comparable in energy use, despite its location and summer use, to schools that are verified or emerging zero energy. It has not yet adopted any solar to provide for its energy needs. So, how far was River Crest below regional cost averages?

Regional school construction cost averages have been published annually by School Planning and Management.[2] These cost averages for elementary, middle schools, and high schools have been useful for estimating conventional school costs for comparison purposes, although some adjustment is required for higher cost localities within the region. Unfortunately, only national averages are available after 2015. The region in River Crest's case is Illinois, Minnesota, and Wisconsin. The Elementary School Cost average for this region in 2008 was $223 per square foot. River Crest is 93,500 square feet and was built at $129 per square foot.[3] That is an extraordinary difference in cost.

As the regional cost average was perhaps influenced by higher costs for schools in and near major cities in the region, such as Chicago, Milwaukee, and the Twin Cities, another comparison is provided by an elementary school built in New Richmond, Wisconsin, which is in a similar setting 20 miles from River Crest. Hillside Elementary School is 85,500 square feet and was completed in 2007. It has some green features but was not LEED certified. The construction cost was $159 per square foot.[4]

The example of River Crest Elementary upsets the conventional economic paradigm. The addition of more performance and infusion of sustainability did not push its cost above or even close to the conventional regional averages. It also had a lower construction cost than a nearby more conventional school in a very comparable setting. This raises interesting questions. Can sustainability with its high performance be included in school designs that are at equal or lower cost than conventional schools? Can sustainability be used to drive school project costs lower? Does an integrated project delivery approach, often used in sustainable projects, drive project costs lower?

A variety of professional organizations have coalesced around the notion that sustainable projects, before considering solar, can and are being built at costs comparable to conventional costs. The recently released ASHRAE (American Society of Heating, Refrigerating and Air-Conditioning Engineers) *Advanced Energy Design Guide for K–12 Schools—Achieving Zero Energy* includes forceful statements on the cost competitiveness of sustainable building with low Source EUI levels.[5] The theme of how to contract and how to construct sustainable schools and provide remodeling at conventional cost or less will be considered later.

The ASHRAE *Advanced Energy Design Guide* notes that sustainable schools with low Source EUI levels are cost competitive with conventional school costs. This is before considering the cost of on-site solar PV. Recent school projects with solar, however, provide evidence that on-site solar can be included and still match conventional school cost. There are two paths to get to this happy outcome. One is to identify cost-saving opportunities in a sustainable school project sufficient to afford purchasing the addition of solar. The second path is to use solar provided by third-party investment.

In the second path with third-party investors, the concept is that an on-site solar PV system is added to a new or existing school with minimal or no upfront cost and the solar power produced will be sufficient to reduce purchased electricity cost and allow for the solar system to be eventually paid off. Alternately, financially advantageous solar can be provided in many states through per power purchase agreements or long-term leases. In these cases, the schools may never own the on-site solar PV systems, but benefit from paying lower costs for solar power than what they would have paid in buying power from the electric utility. This was the case for the Kern High School District Case mentioned in chapter 2.

An example of third-party solar investment is provided by the Darlington Community School District in rural, southwest Wisconsin, where a financially beneficial solar energy system was added to two existing schools in 2016 without the need for up-front funds by the school district. Third-party investors added a single solar PV system that serves the high school and adjacent K–8 school. The school district makes monthly payments through an energy services agreement for solar power and some education and demand management services and has options to buy out the solar system at fair market value at specified future dates.

The reasons for this largely unanticipated opportunity to bring solar to schools on a financially advantageous basis—that is, it reduces energy costs—in the last few years is a combination of dramatic declines of about 75 percent in the last decade in the cost of solar, along with Federal incentives for third-party investors and some financial incentives in some states. What the future holds for solar PV system costs, tax treatment, and incentives will continue to evolve, but the momentum in the market is strongly in favor of a further massive expansion of solar.

Based on the cost experience of schools such as River Crest Elementary School with respect to sustainable design, or of Darlington Community School District and Kern High School District with adding solar energy with third-party investors, the answer to where we want to go can be stated in a simple way. *We want sustainable schools that cost less than the regional conventional cost average and provide a high level of learning performance and energy performance. In other words, we want affordable, sustainable solar schools.*

The design of schools, whether new or being remodeled, is a complex task. Simplifying the discussion to the learning environment and energy is an oversimplification. There are additional objectives in school design and construction:

- Various types of learning spaces and aesthetics
- Flexibility of spaces
- Durability and maintainability
- Daylight and views
- Thermal comfort
- Indoor air quality
- Acoustic performance
- Security

There is a more profound response to the question of where we want to go from the current mixed bag of public and private schools. Where we really want to go is to a high performance learning environment that meets the objectives just listed and others. The energy requirements for such a school will vary with geography.

The ASHRAE *Advanced Energy Design Guide for K–12 Schools* considered 23 climate zones covering the United States and developed Target Source EUI levels for an elementary school and for a larger secondary school.[6] These ranged from 50 to 71 kBtu/ft^2 per year depending on the climate zone. The logic is that if these targets can be approached or achieved, the amount of on-site solar PV capacity to reach zero energy is reduced, making a solar PV system purchase even more affordable. If TPI solar is used, there may not be an up-front cost savings from a reduction in the size of the solar PV system, but the monthly operating cost in purchased solar energy is lowered.

While River Crest Elementary and most of the schools on NBI's zero energy list don't reach the ASHRAE Source EUI targets in their operation, they come close enough. Their energy performance makes it feasible for solar PV to be installed to achieve zero energy and save on their operating cost. At these energy-use levels, the energy required to operate schools could be provided by on-site solar energy in most cases. In cases where space is particularly limited, zero energy can sometimes be provided by a mix of on-site and local off-site solar.

Where we want to be going is for new schools to become zero energy. These zero energy schools will be grid connected, providing power to the grid at times and taking power at others. Batteries may be necessary to store energy to provide power at night or times of insufficient solar production. The batteries will also provide power at times when needed by the grid, but not by the school within the context of smart buildings and a smart grid.

The initial and possibly long-term ownership of the on-site solar and, in some instances, off-site solar may be with third-party investors. The main financial consideration for school districts in solar system ownership is what is the lowest cost approach to providing the on-site solar. Noting that on-site solar is already out-completing grid-based electricity from central power plants in many locations, on-site solar will become an important part of the energy picture for schools.

Thus, the answer to the question of where we want to go moves far beyond a narrow marginal benefit / marginal cost analysis of what features can be added to an existing school remodeling project or be added to a design for a new school. The answer recognizes the examples of recent schools at very competitive total project costs that are zero energy capable—and especially schools that have already included solar to reach zero energy.

The sustainability implications of this response are transformative. Schools, at least in their operation, would no longer be contributing CO_2 to the atmosphere and consequently would become part of the climate-change solution rather than part of the problem. While the details differ, similar opportunities exist for other types of commercial buildings. With respect to on-site solar PV power, many major business entities have already installed substantial solar systems. Firms with more than 50 MW of solar PV as of 2016 include Target, Walmart, Costco, and Apple.[7] It's time for schools to take advantage of this opportunity.

The question is how to get from the current situation to the sustainable, solar-powered school of the future.

NOTES

1. *Getting to Zero Status Update and List of Zero Energy Projects*, New Buildings Institute, January 23, 2018.

2. See, for example, School Planning and Management, July 1, 2015. 20th Annual School Construction Report, February 2015.

3. American School & University, November 2009. Architectural Portfolio 2009, River Crest Elementary School.

4. American School & University, November 2009. Architectural Portfolio 2009, Hillside Elementary School.

5. ASHRAE, *Advanced Energy Design Guide for K–12 School Buildings—Achieving Zero Energy*, Atlanta: ASHRAE, 2018.

6. Ibid.

7. Solar Energy Industries Association, 2017.

Chapter 4

How Do We Span the Divide between Here and There?

A starting point in answering this question is to point out that the divide between current practice in school design and construction and a sustainable solar school is a lot smaller than many people would imagine. After all, there are examples of zero energy schools and other high performance commercial buildings that have been built at competitive costs. And there are many other schools that are zero energy capable but have not yet added on-site solar or sufficient solar capacity required to achieve zero energy. Adding the on-site solar may only be a matter of time and need not require up-front funding by the school district or private school.

A group of friends pondering their first-time trip on the river they've chosen into a remote area might ask similar questions. Can we get over the trip planning and commitment hurdles? Once the trip is launched, can we overcome obstacles during the trip? It's something they are attracted to do, they have some or even all the equipment for the trip including the maps and guide books, and they have most or even all the skills and experience required. At a finer-grained level, there are always some unknowns—perhaps about some of the rapids or portages, how the weather will be, what the actual water levels will be, and who finally signs on to do the trip.

If the trip is ultimately going to happen, the group must come to a decision and make the commitment to go ahead. Once a decision has been made, further detailed planning takes place in terms of specific dates, travel arrangements including arranging for vehicle shuttles, drop offs and/or pick-ups, and so on. There's always a chance that something comes up in the detailed planning and the trip gets called off.

Once the trip has started and you've left the departure point on a river, at least, you are committed. This is similar to the point in a new school project or major remodeling that construction drawings are completed, the construction work has been bid and contracted, and the ground has been broken. There's no turning back. There may be unanticipated adjustments after departure on the river such as in the schedule, but you aren't turning back and paddling up rapids back to the starting point.

Only in rare circumstances will some commercial building projects get stopped during construction by something like a bankruptcy on the part of the owner, which is almost unheard of in the case of public schools. This is equivalent to a catastrophic accident on a river trip, such as having a raft destroyed in a rapids or major injuries. At that point, rescue is required. Helicopters are generally not allowed in the Grand Canyon National Park. One exception is for accidents, where injured people require air rescue.

A clear vision of where you want to go and a firm commitment to that goal is required before a group gets too far into the serious planning for a river trip. For a school district, however, the situation is only slightly different in that there may be some flexibility in the destination even after a school project has started construction.

For example, a school district may decide to take the path of a zero energy capable sustainable building that meets all its goals for space and as a learning environment. The Source EUI is in the range of 40 to 120 kBtu/ft^2 per year. The solar PV plan is to have rooftop solar provided through third-party investors using a power purchase agreement (PPA). If during the construction the school district can't arrive at satisfactory terms on the PPA, the school may have to change course and hold off on installing some or all the planned solar. In this case, the school district will still have an exceptional school with lower energy costs. In the future, a suitable PPA arrangement may be found to allow some or all the solar to be added.

The New Buildings Institute (NBI) is an important catalyst in driving better energy performance in commercial buildings in general and in facilitating zero energy buildings. As stated on its website, *"NBI has been leading the market development of zero energy (ZE) buildings since 2008 when we supported the development of the first ZNE Action Plan to help California meet its ambitious zero energy goals."*[1] NBI organizes research and training on zero energy buildings and maintain lists of buildings that have gone through a certification process for zero energy and of buildings that are in the process of getting to zero energy.[2]

As of January 2018, there were 67 zero energy verified buildings and another 415 emerging buildings—that is, buildings that plan to attain zero energy verification. There are thousands of additional buildings across the country that are performing at energy-use levels that readily support zero energy if solar or other renewable energy were added to the site.

This assertion that there are thousands of existing buildings that are capable of zero energy can be applied to schools. This is evident noting the range of the Source EUI for the schools on the NBI zero energy list and comparing that to the range of the known Source EUI for existing schools provided by Energy Star Portfolio Manager. As mentioned in chapter 3, the range of Source EUIs for schools in NBI's list was from 48.8 to 160.7 kBtu/ft² per year. The national median Source EUI for schools in the EPA Portfolio Manager is 114. With 55,000 schools in the database, this implies that 27,500 schools might be considered to be at energy-use levels that would support zero energy if one used the median value of 114 Btu/ft² per year as the upper limit for consideration. The number of schools that might be considered for zero energy increases as the upper limit threshold value is increased.

How many of these schools have considered at least some solar and have plans to add solar is unknown. Given the precipitous decline in solar PV costs since 2008 and rapid expansion of solar installations and installed capacity in schools, it's likely that many of these schools will be considering solar in the next few years.

Installed solar PV system costs in the upper Midwest, for example, declined from more than $8 per watt in 2008 to under $2 per watt in 2018 for system sizes in the range of 200 to 800 kW, suitable for most schools. A study undertaken in part by scientists at the National Renewable Energy Laboratory (NREL) documents how the costliest part of the PV system—namely, the price of the PV modules or panels—have declined over time as production has increased as shown in figure 4.1.[3]

The decline in module prices and the increase in cumulative module production has been so enormous that logarithmic scales are used. The figure

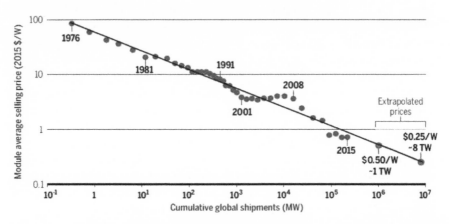

Figure 4.1   **Declining Solar PV Panel (module) Price.** *Source*: Haegel et al. 14 April 2017. Terawatt-scale photovoltaics: Trajectories and Challenges. Science 356 (6334). GRAPHIC: N CAREY/SCIENCE.

reveals how module prices that were near \$10 per watt in 1990, had fallen below \$1 per watt in 2012, and were extrapolated to be near \$0.5 per watt by around 2020. The precipitous solar panel cost decline and overall installed solar PV system cost decline explain why solar PV systems have so suddenly emerged as a financially preferred option for schools.

The comparison of energy use in schools and what level of energy use might be the threshold to consider zero energy using solar suggests two lessons. The first is that there is room for further efficiency improvements, especially for schools with above the median level of energy use. Making these improvements will reduce operating costs and better position the school for adding on-site solar. Energy efficiency investments should be made until the point where the cost per dollar of energy saved of further efficiency investment is equal to the cost of on-site solar per dollar of energy produced. Then solar takes over. The operating principle for a school is that investments in efficiency and solar should be undertaken when financially advantageous for the school.

The second lesson, especially for schools that are near or below the median Source EUI energy-use level is that the divide between those schools and the zero energy schools of the future is largely attributable to lack of information and other implementation barriers in the solar market. After financially advantageous energy efficiency investments are accounted for, attention needs to focus on solar market and financing issues to cross the divide to solar zero energy schools.

The current challenges for on-site solar include:

- Unfavorable utility regulatory environments with respect to solar in some states. These include limitations such as low net metering limits and prohibition of PPAs (power purchase agreements) and other TPI options.
- Lack of battery or other energy storage systems at lower costs levels that would be provided by mass production.
- Electric utility grids that are not sufficiently smart and are generally ill prepared for a world with large quantities of on-site renewable capacity along with large quantities of remotely sited renewables from wind farms and solar farms. Some locations in distribution grids cannot absorb all the excess solar production a school may be able to produce at certain times.
- The predominance of natural gas for space heating in most schools. Using on-site solar to achieve zero energy in a school using natural gas for heating requires a net annual export of electricity, which exacerbates the issues of absorbing power from on-site solar systems on the grid.

Interestingly, the ability of on-site solar to compete with the cost of grid-provided electricity is not included in this list of barriers in spanning the divide to zero energy schools. In most locations in the United States, solar costs are equal to or lower than centrally provided power. In the meantime, there are two other challenges that are often presumed to contribute to the divide between today's schools and where we want our schools to be in the future:

- Insufficient roof and ground space is available for siting on-site solar.
- The solar market is not sufficiently developed to meet a rapidly rising demand.

These barriers are more perception than reality.

Let's consider these challenges beginning with the last two. Schools tend to have lots of roof space. A common characteristic of most of school construction in the United States is the predominance of single-story construction. The consequence of this construction pattern is the availability of extensive roof area that mostly provides no other value than serving as a roof, with a small portion providing space to locate mechanical equipment. This mostly unused area can be turned into a resource for harvesting solar energy. The question is this: How does the available roof area compare to the area required to provide solar energy to serve all the needs of a school?

The ASHRAE *Advanced Energy Design Guide for K–12 School Buildings—Achieving Zero Energy* looked at this question in considerable detail.[4] The approach was to use models to simulate energy use of an 82,500 square foot prototypical elementary school following their design and construction guidelines in each of the ASHRAE climate zones. It should be noted that the modeling assumed the guidelines were fully implemented, resulting in a very high performance school. The finding was that in the case of single-story schools, from 16 to 31 percent of the roof area was required for solar except in the case of zone 8, which is subarctic and arctic, where 45 percent of the roof area was required.

Stepping back from the recent ASHRAE *Advanced Energy Design Guide*, consider the case of a single-story, 93,500 square foot River Crest Elementary School, referenced in chapter 3. Completed in 2008, it does not match up in energy performance to the ASHRAE *Design Guide* in part because it has year-round operation. Its Source EUI is 110 kBtu/ft^2 per year. Its annual electricity requirement was 604,000 kWh. The roof area required to provide that amount of electricity is about 50,000 square feet. River Crest uses almost 39,000 therms of natural gas for heating, domestic hot water, and the kitchen. Providing solar to offset that natural gas use would require another 32,000 square feet of roof area for a total of 82,000 square feet of the 93,500 square feet of roof. Thus, even in a very cold climate with hot summers, and with extensive summer use, there is sufficient roof area to support solar.

The conclusion from considering the ASHRAE analysis of optimized buildings and the River Crest example is that schools provide large roof areas that will often be sufficient to locate on-site solar.

There are variants to be considered, the foremost of which is the case of a two-story or multistory school, which presents an obvious challenge for roof area. Some schools, especially in urban areas, may have shading from adjacent buildings and trees. Conversely, schools that don't have air-conditioning and schools that aren't used in summer reduce energy requirements and roof area needed for solar energy. The multistory and/or partially shaded schools are the primary challenge.

There are three possibilities to expand on-site solar beyond the school roof. The first is the availability of open land area not required for athletic fields or outdoor learning space. This opportunity exists on some sites, especially in rural areas. An advantage of ground-mounted solar in areas with substantial snow cover is the ability to use higher angles for the solar panels that provide for better melting and shedding of snow. Solar applications on flat roofs are generally done at low angles such as 10 degrees to avoid wind loading and snow drifting.

Another possibility is parking areas. Solar covered parking as used in the Kern High School District in Bakersfield, California is advantageous as it provides shelter from hot sun as well as from rain and snow, even if weather-proof shelter is not provided.[5] As heat shortens battery life, shading provided by solar covered parking will be of increasing advantage as the number of electric and hybrid vehicles increases.

The third possibility for expanding solar is to use adjacent or nearby land or roof areas such as in community solar development. Depending on the situation, the solar PV system may not feed directly into the school behind the meter but might be fed into the local distribution grid. A few states already allow virtual net metering, which means that the school's PV system can provide power and receive the financial savings under net metering rules by simply putting the power onto the local grid.

The possibilities to expand solar beyond the roof can be used in whatever combination is most appropriate for the school. The recent cost experience with these siting options for solar systems of 200 kW to 600 kW, which is appropriate for many schools, is that roof and ground-mounted PV systems have similar costs. The cost of the parking shading structures, however, increases the installed cost of solar per kW by up to 35 percent but provides the added benefit of shading.

The bottom line is that most existing schools provide conditions suitable for substantial on-site solar PV systems, and many of these schools have sufficient space to add solar to reach zero energy. Some limitations will be found with older and/or multistory schools, especially in urban areas with small sites.

Any questions about the ability of the market to provide and install solar systems is being answered in most areas of the country by the rapid expansion of the solar market. Data tracked by the Solar Energy Industries Association (SEIA) and Wood McKenzie Power & Renewables demonstrate the massive growth in the solar market in the United States, which now employs about 250,000 workers.[6] This phenomenal growth in installed solar capacity was propelled in the early stages by the massive support for solar, especially in Germany, and subsequently by the gigantic expansion of the global manufacturing market led by China.

The SEIA/Wood McKenzie Power & Renewables report shows negligible annual installation of solar PV capacity in the year 2000. By 2010, the total annual installation increased to roughly 1,000 MW (or 1,000,000 kW). It doubled to 2,000 MW annually in 2011. By 2015 it was approaching 8,000 MW. Estimates for 2018 are for the annual installation of PV capacity to be about 11,000 MW and account for about 30 percent of all new electric capacity added in the United States. School solar installations are considered nonresidential installations, larger than residential solar and smaller than utility scale solar. More than 2,000 MW of nonresidential solar was being installed in 2018.

As in any industry in early to mid-stages of development, some turbulence is to be expected, with new entrants, early stage failures, and consolidations. The solar import tariffs imposed by the United States in January 2018 initially caused approximately a 20 percent increase in installed system costs in the United States, but subsequent developments appear to be returning installed costs in the United States toward pre-tariff levels. Specific events that cause upset in the market are unpredictable. As will be discussed later, changes in the solar PV market will continue, but the long-term downward trend in installed solar PV system costs is also likely to continue. This expectation for the PV panels is noted in figure 4.1.

Unfavorable utility rate structures and regulations with respect to customer or third-party-owned solar PV and other renewable energy systems are a challenge in some states, and these vary greatly from utility to utility and state to state. The point of contention is that some electric utility companies are attempting to capture as much electricity production (including solar) and/or sales as possible to maximize profits. Utility rates that include high fixed charges, low buy-back prices when excess solar production from customers is sold back to the grid, and various restrictions on third-party ownership are among the measures that inhibit on-site solar.

The fight for market share is taking place in a dumb environment. As will be considered later, the dumb utility pricing market is serving to somewhat constrain the school solar market both in terms of the number of schools with solar and the size of solar systems being installed. The adjective *dumb* is meant to describe much of our current situation where essential, real-time information regarding the price of electricity supply at any moment is not being provided to electricity users.

The underlying real price of power from the grid is a function of the costs of electricity production, transmission, and distribution, as well as aggregate customer demand at any given moment in time. Most utilities do not yet provide real-time pricing based on cost of service to most of their customers. Hence, customers, including schools, have not made the arrangements for their operating systems to respond to real-time pricing. Schools and other customers get the utility bill at the end of the month, usually give it little if any thought, and pay the bill.

Even in the absence of real-time pricing, some schools are getting smarter as they develop capabilities to limit their peak demand measured in kW (kilowatts). Most schools are billed for their electricity in two parts. The first part is for electricity use measured in kWh and the other part of the bill is based on demand, or the peak monthly kW load being put on the system. The peak kW is measured every 15 minutes. The highest peak kW interval is the basis for the monthly demand charge. The monthly electric utility bill is based on a fixed rate structure or tariff for both electricity use and demand.

For example, there may be an on-peak rate from 9 a.m. to 9 p.m. and an off-peak rate from 9 p.m. to 9 a.m. The rate structure may have higher charges for both kWh purchased on-peak and for peak kW that occurs on-peak during the month. At the end of the month, the bill comes based on how many kWh and kW where used and when.

Reducing both kWh and kW reduce the bill. Even if the kWh use has been reduced to a practical minimum, limiting the peak demand provides further cost savings. The demand savings can be achieved by taking information in real time from the electric meter and transferring it to the building controls system (BCS). When predetermined kW thresholds are approached, the BCS can be programmed to prevent the school from exceeding a target threshold. For example, the BCS might reset some temperatures and/or limit the chiller capacity for a short period of time. The objective is to save the school cost without reducing occupant comfort, and perhaps without even having the occupants noticing that operations have been adjusted on a short-term basis.

Trends in the marketplace are for greater monitoring capability of school building performance and anticipating smart grids with real-time pricing rather than fixed tariff schedules. Smart schools will utilize the price signals to manage energy use and peak demand to the fullest extent possible.

As more lower-cost renewable energy is added to the electrical grid by utility customers like schools and by utilities themselves, a central challenge and opportunity for the electric grid, utilities, and customers is energy storage. Getting across the divide from today's power system with a larger fossil component to tomorrow's low fossil or no fossil component will require energy storage to avoid excessive investment in peak-load power plants—especially when neither sun, wind, and hydro power are available in sufficient amounts. Batteries are expected to be an important part of the future for providing energy storage. Depending on the state and location, batteries will likely be a complement to solar PV systems at many schools.

Battery development is a major locus of research and massive investments are underway in battery manufacturing—for example, at Tesla's manufacturing facility in Reno, Nevada. Chinese and other international manufacturers are also entering the market as they have done for solar panels and inverters. The scaling up of battery production will continue to reduce battery prices for storage while the penetration of smart grids and smart buildings will provide the ability for customers to take advantage of the financial rewards of owning batteries.

The vision for battery operation is not limited to simply storing excess solar energy production from one's own building to be used overnight or at some future time. The operation of batteries in smart buildings connected to a smart grid provides power to the grid based on grid power prices and storage costs for whomever needs it at a given moment. If the grid power sale price overnight is low, batteries could also be used by a school to buy and store low-priced grid power for use in the building during high-cost periods or for later sale to the grid. Batteries can also generate revenue for grid support services including voltage and frequency support. Battery installations as part of solar installations are now emerging in the marketplace. The trends in pricing are anticipated to follow the path of solar PV panels and cell phones.

The final condition that contributes to the divide between current school design and the sustainable, solar-powered school of the future is the natural gas market. After a long period of substantial price fluctuations and occasional shortages, natural gas market prices sharply declined in late 2008 and into 2009 and have remained at historically low prices. An important cause for the low prices has been the increase in natural gas production with the development of hydraulic fracturing or fracking, which has opened new production areas and formations at an affordable cost—if one ignores the environmental costs or externalities. Natural gas pipelines have also expanded in this period to service expanded market areas, including international sales.

For most schools with substantial heating requirements, natural gas has provided a low-cost heating source for the last decade. For schools in the parts of the country with substantial heating loads, like the Northern Great Plains, the Rockies, the upper Midwest, and parts of the Northeast, natural gas typically accounts for 20 to 40 percent of a school's total energy bill. This use of natural gas presents a challenge to achieving zero energy schools with respect to achieving zero energy when schools are also focusing on minimizing initial project cost.

There are two possible paths for a school district considering a zero energy school where there is significant heating need and natural gas is available:

1. Utilize natural gas and over-size the on-site solar to be able to export sufficient solar electricity to the grid to offset the carbon content of natural gas use.
2. Forgo natural gas in favor of electric heat. With electric resistance heating an inefficient choice, if it's even allowed by code, the far more efficient approach would be to use heat pumps with their added first cost. As noted in chapter 2, *geothermal* is the term commonly used for heat pump systems that exchange heat with the earth and will be the most efficient option for an all-electric school. Less costly air-sourced heat pumps are a less efficient but may be an effective option for climates that are not too cold. Progress is being made in improving the efficiency of air-sourced heat pumps for cold climates.

From a sustainability perspective, the option to use natural gas is problematic for several reasons. The use of natural gas results in carbon releases into the atmosphere. In addition, a small but highly consequential amount of natural gas is released in the form of methane leaks from the production and transportation process. The impact of methane as an atmospheric, heat-trapping gas per molecule is 25 times worse than CO_2 emissions. Despite this, natural gas is often considered a transition fuel or bridge to a carbon-free energy system because it releases about half the carbon per kWh of power production relative to coal.

Additional environmental issues from natural gas include other consequences related to fracking including groundwater contamination from fracking fluids, the components of which do not need to be reported under U.S. law. Thus, although natural gas has lower carbon content than coal–based electricity, its continued use is not sustainable.

Natural gas presents a further quandary for a school district considering a long-term investment in a new school or a major remodeling. For how long will natural gas provide a competitive-cost energy source considering the high costs of changing a school's mechanical systems in the future? A commitment to natural gas in a new school is at least a 25-year decision, and likely much longer. Can natural gas prices be counted on to stay low for 25-plus years, especially as natural gas producers expand export capability using LNG (liquified natural gas) terminals? International sales will serve to increase U.S. natural gas prices.

National carbon pricing is a potential development that would also increase the cost of natural gas relative to solar power. The United States is likely to be joining other regions of the world and California in adopting carbon pricing in some form—either a carbon tax or a cap-and-trade system. The United States previously adopted a cap-and-trade system to effectively manage leaded gasoline phasedown and the European Union has a large Emission Trading System for carbon and other greenhouse gas emissions.[7]

To get to the other side of the divide between where we are and want to go, schools would avoid natural gas for new projects in anticipation that the initial or first cost savings in a school project using gas-fired HVAC systems relative to geothermal systems may be given back if natural gas costs increase due to the finite nature of natural gas supplies. Environmental problems are making natural gas increasingly unacceptable.

Whether a school district is financially ahead or behind in the near term in going the all-electric route using heat pumps depends, especially in areas with higher electricity tariffs, on whether it is choosing solar PV systems to meet all or most of its energy needs. Solar power provides the source of lower-cost power for geothermal systems to out-compete natural gas HVAC systems. Heat pumps systems may not financially compete in an area of very high electric rates and low natural gas rates without on-site solar. Thus, schools located in high electric cost areas should be particularly motivated to adopt on-site solar.

The remaining divide to the sustainable, zero energy solar school of the future is small and diminishing. The use of third-party investors is an attractive choice to relieve school districts of the first cost barrier. The emergence of intelligent buildings and grids is facilitating the transition as schools can better financially manage on-site electricity use, solar production, and storage.

Thomas Edison had a prescient comment on the use of oil and coal, and certainly would have included natural gas had that been in the marketplace at the time of his comment: *"I'd put my money on the sun and solar energy. What a source of power! I hope we don't have to wait till oil and coal run out to tackle that."*[8]

NOTES

1. "Zero Energy," New Buildings Institute, https://newbuildings.org/hubs/zero -net-energy/.

2. *Getting to Zero Status Update and List of Zero Energy Projects*, New Buildings Institute, January 23, 2018.

3. Nancy M., Haegel, Robert Margolis, Tonio Buonassisi, David Feldman, Armin Froitzheim, Raffi Garabedian, Martin Green, Stefan Glunz, Hans-Marin Henning, Burkhard Holder, Izumi Kaizuka, Benjamin Kroposki, Koji Matsubara, Shigeru Niki, Keiichiro Sakurai, Roland A. Schindler, William Tumas, Eicke R. Weber, Gregory Wilson, Michael Woodhouse, and Sara Kurtz, "Terawatt-Scale Photovoltaics:

Trajectories and Challenges: Coordinating Technology, Policy, and Business Innovations," *Science:* Vol. 356, Issue 6334, April 14, 2017.

4. ASHRAE, *Advanced Energy Design Guide for K–12 School Buildings—Achieving Zero Energy*, Atlanta: ASHRAE, 2018.

5. www.sagerenew.com.

6. "U.S. Solar Market Insight Report for Q3," www.SEIA.com, from Wood Mackenzie Power & Renewables and the Solar Energy Industries Association, September 2018.

7. Robert M. Stavins, "What Can an Economist Possibly Say about Climate Change?," November 7, 2014, in *The Annual Proceedings of the Wealth and Well-Being of Nations 2014–2015*, Vol. VII: *Economic Policy and the Challenges of Climate Change Ideas and Influence of Robert N. Stavins*, ed. Warren Bruce Palmer, Beloit, WI: Beloit College Press. 2015.

8. James Newton, *Uncommon Friends: Life with Thomas Edison, Henry Ford, Harvey Firestone, Alexis Carrel, and Charles Lindbergh*, San Diego: Harcourt, 1987, 37.

Chapter 5

The Inevitability of Solar Schools

The scientific literature on climate change is extensive and rapidly growing. There is a broad consensus that human activity is the dominant cause of climate change for the period from 1951 to 2010.[1] The Intergovernmental Panel on Climate Change (IPCC) has served a central role in gathering, assessing, and disseminating the international work on this subject. The IPCC Working Group II concluded that severe, pervasive, and irreversible impacts are likely to occur when global average temperatures increase by more than 2 degrees centigrade.[2]

A blunt way of summarizing this body of work would be to say that we are in a desperate situation for life on earth, including the survival of the human species. We've taken a lot of carbon stored in the earth in fossil fuels and released that in the form of CO_2 into the atmosphere and into the oceans. It will take decades for the climate change consequences from the CO_2 already emitted along with other greenhouse gases to play out. Substantial further warming beyond the 1-degree centigrade increase already experienced is baked into our predicament.

The IPCC review of the science makes clear that reversing the trends in global warming is not in the cards. There was a time from about 1980 to 1990 when we had sufficient understanding of climate change and when policies to better limit some climate change trends might have succeeded, but that ship has sailed.[3]

What is in play by this late time in the crisis is limiting the amount of warming to levels that can somehow be managed as our species and the natural ecosystems on which we depend adapt to climate change. Key systems on which we depend include agricultural systems and our hydrological systems for water supply. Changing rainfall and temperature patterns portend severe challenges. Limiting sea level rise is a paramount concern to the 600 million plus of our global citizens who currently live less than 30 feet above sea level.[4] Some heroic geoengineering schemes have been proffered.[5] And perhaps we'll have to pursue some of those.

The concept of inevitability can be thought of in two different ways. One approach, let's call it *forced inevitability*, would be to ask what must happen because of events that give us no choice—that is, we've got to do it. The Ebola outbreak in 2014 caused a global panic and certain inevitable, mandatory responses including closed borders, travel bans, and the imposition of a set of medical protocols.

A different approach to inevitability, let's call it the *voluntary economic response*, is to recognize specific scientific discoveries and technological innovations and how they lead to inevitable economic choices and developments. The developments are so economically superior that competing technologies, systems, or firms don't have a chance. Innovation in computers and fiber optics enabled the emergence of the Internet. Some of the inevitable voluntary responses have been the emergence of online commerce with firms such as Amazon rapidly displacing segments of retail business. A century earlier, the discovery of oil and the invention of the internal combustion engine ended the use of horses in the U.S. transportation system. The mechanisms were again voluntary economic forces.

Buildings powered by on-site solar PV systems may be an example of a development that is inevitable, both as a voluntary economic response that is superior to the status quo, as well as a policy that might be forced by local, state, national, and international responses to climate change and its drastic consequences. A forced path to inevitability in the form of schools required to adopt substantial or zero energy levels of solar would make no sense if restricted to schools. Such a policy would be part of broader policies that would accelerate the transition to renewable energy and away from fossil fuels in buildings, agriculture, industry, and transportation.

The work of the IPCC panel describes mitigation scenarios for an atmospheric CO_2 concentration of about 450 ppm in 2100.[6] Limiting CO_2 to that level is associated with a global temperature increase of 2 degrees centigrade. Such scenarios include large energy-use reductions in buildings as well as a 90 percent reduction relative to the year 2010 in CO_2 emission by the energy-supply sector in the years 2040 to 2070. Solar PV, whether situated on schools or elsewhere, would be the type of energy supply that would support such mitigation scenarios.

Whether this happens primarily due to market forces or in part due to the force of regulation cannot be known. If voluntary forces aren't materializing, regulatory approaches may be the response to a desperate, all-hands-on-deck situation. Thus, zero energy buildings and buildings approaching zero energy may be inevitable in the not too distant future. The technology exists now, and the market forces have already shifted, so we might as well commit to this path now.

As described in chapter 4, there are some schools that are already zero energy, and at least 27,500 schools whose energy requirements are less than a Source EUI of 114 kBtu/ft^2 per year and thus would make them candidates to get to zero energy with the addition, or in some cases expansion, of on-site solar and on-site energy storage. Additional energy-efficiency measures when less than the cost of added solar would also be involved. The case could be made that a robust smart grid with centralized energy storage could suffice without on-site batteries, but the argument for some on-site storage being part of this future seems more credible.

A voluntary effort within the architecture and construction community and other supporting entities for zero energy buildings is the 2030 Challenge.[7] The initiative calls for the phased reduction in fossil fuel required in the operation of new buildings, developments, and major renovations by the following amounts:

- 80 percent in 2020
- 90 percent in 2025
- Carbon neutral (i.e., 100 percent) in 2030

The 2030 Challenge promotes the use of on-site renewables, with a 20 percent maximum for renewables located off-site. The 2030 Challenge's website states that 80 percent of the top 10 architecture/engineering/planning firms in the United States have adopted the challenge. The AIA, ASHRAE, and various governmental organizations have adopted the challenge along with the Royal Architectural Institute of Canada. As of early 2018, 462 firms in the AIA had already joined the AIA 2030 Commitment.

The 2030 Challenge is a response to the question of what mix of market forces and regulation makes the solar school inevitable. It's an inspiring example of the voluntary path. While firms and organizations are adopting the challenge voluntarily, the state of California has signaled that it does not want to take chances on reaching zero energy quickly. It is making zero energy a requirement for new residential buildings in 2020 and for new commercial buildings in 2030.[8] The segment of the building market being addressed by such regulations is not the innovators and early adopters of zero energy that are already voluntarily moving, but the market laggards.

Voluntary forces pushed by the market may be nimbler for the economy and would be the emerging path that most in our society would prefer, including school districts and private schools. How much additional impetus by force, and hence via regulation, is needed will be determined by events. In any event, solar schools, many of them zero energy, are coming your way soon.

NOTES

1. T. F. Stocker, D. Qin, G. K. Plattner, M. Tignor, S. K. Allen, J. Boschung, A. Nauels, Y. Xia, V. Bex, and P. M. Midgley, eds., *Climate Change 2013: The Physical Science Basis—Working Group I Contribution to the Fifth Assessment Report of the Intergovernmental Panel on Climate Change: Summary for Policy Makers*, Cambridge and New York: Cambridge University Press, 2013. Available online at the Intergovernmental Panel on Climate Change website, https://www.ipcc.ch/report/ar5/wg1/.

2. C. B. Field, V. R. Barros, D. J. Dokken, K. J. Mach, M. D. Mastrandrea, T. E. Bilir, M. Chatterjee, K. L. Evi, Y. O. Estrada, R. C. Genova, B. Girma, E. S. Kissel, A. N. Levy, S. MacCracken, P. R. Mastrandrea, and L. L. White, eds., *Climate Change 2014: Impacts, Adaptation, and Vulnerability—Contribution of Working Group II to the IPCC's Fifth Assessment Report*, Cambridge and New York: Cambridge University Press, 2014. Available online at the website, Intergovernmental Panel on Climate Change, https://www.ipcc.ch/report/ar5/wg2/.

3. Nathaniel Rich, "Losing Earth: The Decade We Almost Stopped Climate Change. A Tragedy in Two Parts," *New York Times Magazine*, August 5, 2018.

4. Deborah Balk, Institute for Demographic Research CUNY, 2018.

5. Alan Robock, Luke Oman, and Georgiy L. Stenchikov, "Regional Climate Responses to Geoengineering with Tropical and Arctic SO2 Injections," *Journal of Geophysical Research Atmospheres*, August 16, 2008.

6. "Summary for Policy Makers," in *Climate Change 2014: Mitigation of Climate Change. Contribution of Working Group III to the IPCC's Fifth Assessment Report*, 2014. Available at the Intergovernmental Panel on Climate Change website, https://www.ipcc.ch/report/ar5/wg3/.

7. Architecture2030.org.

8. CPUC.ca.gov, California Energy Efficiency Strategic Plan/Zero Net Energy.

Section 2

DEPARTURE

Chapter 6

So, You Want a Sustainable Solar School

Starting out on the path to a solar school has similarities to the initial planning for a river trip. The starting point of a river trip is a choice of what river you want to travel, including the beginning and ending points, and who is going to be on the trip. The subsequent decisions flow from that starting point. Someone needs to choose the time of year, types of boats, safety and first aid equipment, travel pace, and mix of activities such as fishing, side hikes, and photography. Food, beverages, cooking gear, kitchen ware, tents, waste management, and cleaning equipment need to be figured out and provided.

There are many elements that need to come together to get, say, a two-week trip planned, launched, the river run hopefully with good cheer, and successfully and safely concluded. Not the least of the elements is assembling a team that wants to make the trip.

The path to a new school is a longer path and much needs to come together. There is generally some set of needs—perhaps overcrowding or a failing, inadequate facility that will require substantial investments merely to keep the facility operating. In most locations in the United States, a new school is only going to be built if the school district community can sufficiently get behind the new school to support a referendum to borrow money in the form of bonds. A private school has a similar challenge of generating support from its community and raising the funds required.

The question is where to begin the planning, design, and construction process.

As in the case of the river trip, assembling a good team is essential. In most cases, the initial part of the school project team will be assembled in the planning prior to and leading up to the referendum seeking citizen approval for public schools and prior to fund-raising for private schools. Once a referendum has been passed for public schools or fund-raising has succeeded

49

for private schools, the project team will be filled out. Members of the team include the following:

- Planners
- Architects
- Landscape architects
- Interior designers
- MEP (mechanical, electrical, and plumbing) engineers
- IT (information technology) engineers
- Civil and structural engineers
- Solar PV system designers and installers
- Energy modelers
- Construction managers
- Commissioning agent

These team members will be hired for their experience in sustainable school facilities and experience in working as teams, possibly in what's call *integrated project delivery*. Integrated project delivery is a process that coordinates the design and construction effort from the earliest stages of the project. This differs from the design-bid-build process where the design is developed without detailed input from the construction managers who are added to the team later when the general contractor and other specialty contractors are selected.

Whatever the delivery method, the professional team must get in synch with the school administrators, facilities managers, teachers, and school board, and with each other to begin the process of working as a team.

If teamwork and interpersonal dynamics are not highly determined before the river launch, things usually get worked out in the early going, possibly with the help of a few arguments. This may also be the case in the school project albeit with the understanding that the school district administration and school board are the client and ultimate decision makers. The task of the team of professionals is to provide the best information and guidance possible to empower and enable the school district administrators, staff, and boards to make informed decisions.

A river trip has a predetermined destination, and a successful trip gets everyone to the destination, hopefully with fond memories and stories to tell. A zero energy capable or fully zero energy solar school project also has a predetermined destination. The destination is defined in the early planning with careful programming with input from teachers, students, parents, and administrators. A successful school project includes, by the end, a completed, fully operational school on budget and on schedule. Perhaps the best evaluation of a successful project comes from the initial and ongoing reaction of the students, teachers, and community members the first time they enter the new school.

Northland Pines High School in Eagle River, Wisconsin, is a case in point. Northland Pines was one of the early, sustainable schools in the United States, becoming the first public high school in the country to be certified at LEED Gold in 2007. The teachers were stunned when they first entered the school, which was a dramatic contrast to the physically failing school that was being deconstructed. The student reaction was similarly dramatic. The testament to the difference of the new sustainable school was how, over the coming months and then years, the graffiti that had marred the old school never reappeared at the new school.

While the design details of a new school cannot be predetermined at the beginning of a new school project, something like the following would be set as building performance requirements at the beginning of the project:

- A lower Source EUI to enable the school to reach zero energy initially or in the future
- An evidence-based design process guided by energy modeling
- On-site solar PV to provide a significant share of or all the energy requirement
- Water savings of at least 40 percent relative to code
- Construction cost at or below the conventional cost average for the region
- High indoor environmental quality, including considerable daylight, views, indoor air quality for CO_2 to remain below roughly 1,100 ppm, and good acoustic performance including low background HVAC noise levels
- Commissioning required of all MEP systems and the building envelope
- Relative ease of maintenance

Requirements like these might seem obvious elements in an RFP (request for proposal) for professional services for a new school or a major remodeling project, but they rarely are included as a package in RFPs for planning, design, and construction services, preferably with an integrated project delivery approach to include knowledge from the construction professionals from the beginning of the project. The reasons that requirements such as these don't appear in RFPs is perhaps for two overarching reasons: tradition and hesitant or underinformed buyers.

A simplified description of tradition in the design and construction market place is for a small, architect-led design team to go through a planning process with the school district or private school. After determining the needs and desires for the school, preliminary designs emerge for review by the school district. At some early point, preliminary construction estimates are provided, with limited input from construction professionals. Eventually, a budget number is set, and approval is sought through a referendum process or, depending on the state, other budget approval processes. For private schools, the question becomes whether donors commit to the project.

If the referendum passes, the design team is assembled, often from the existing pre-referendum team or through a subsequent competitive process. After the design is completed, the construction is put out to bid. It is only after the construction contractors are selected that the full project team is in place. It's common to have to alter designs at the end of the design work to meet budget. This is referred to as *value engineering*, although some refer to this revision process as design assassination.

The challenge in designing and building to a specific referendum amount is that construction costs cannot be precisely known until a complete set of construction bids has been received. Material and equipment choices and costs are continually changing, as are competitive conditions in the construction market.

The revisions to make budget can result is some unhappy compromises. Having early construction input with more accurate cost estimates and creative interactions on constructability is useful in general, and especially in minimizing late-stage value engineering.[1] Integrated project delivery is particularly effective in incorporating construction manager expertise.

The other reason a list of requirements such as those listed above rarely appears in school RFPs is that buyers often don't know that they can and should make that ask. District administrators, facilities personnel, business managers, teachers, and boards often don't have the background information and experience to realize that they can develop a specific list of requirements and ask for it at the beginning of their school project, when they are assembling their project team. And if the school districts don't ask for these outcomes at the outset, what are the chances that they'll achieve these outcomes?

Discovery Elementary School in Arlington, Virginia, is a 98,588-square-foot, zero energy school completed in 2015. Except for the zero energy goal, the design team worked with the owner at a very early stage to develop requirements like this list. It is an example of a new school that achieved zero energy in performance but did not initially realize zero energy was possible within the budget.[2] The project team played an important role in guiding the school district to this outcome. It is also in the LEED certification process at the platinum level. We'll return to this remarkable school later as a case study.

The point here is that solar schools should be the result of deliberate asks at the beginning of the school procurement process and not the unlikely confluence of late-stage decisions, an excellent team, and fortuitous events.

There are some programs and rating systems that have been developed by the professional building community since 2000 that provide useful tools that are supportive of zero energy capable schools and other commercial buildings. The recently released ASHRAE *Advanced Energy Design Guide for K–12 School Buildings—Achieving Zero Energy*[3] and the *Living Building Challenge*'s Zero Energy Building (ZEB) Certification[4] are excellent sources. Neither, however, is explicit on setting a financial goal of meeting conventional construction cost while including the on-site solar.

The LEED for Schools Rating System is well established and has a strong presence in the United States and to some extent globally.[5] The certification process is a useable and useful system to address many areas of sustainability in a school and other types of buildings. Zero energy and other attributes of sustainability can be achieved, of course, without formal third-party certification, although this is not encouraged by the USGBC. This is understandable, as it's tempting to cut corners and not meet all the requirements for the LEED prerequisites and the credit areas being pursued.

A current limitation of the LEED rating system as it applies to a zero energy sustainable school is that it does not recognize more than a 42 percent improvement in its scoring for energy performance relative to the LEED baseline including on-site solar. Similarly, in terms of credits, it does not recognize on-site renewables providing more than 15 percent of the school's energy use. Thus, LEED does not reward going from 15 percent solar all the way to zero energy or beyond. This omission will perhaps be rectified as more zero energy solar schools emerge.

The *Living Building Challenge*'s Living Building Certification is the most demanding sustainability certification. It also offers petal certification covering seven areas or petals. Certification is based on audits of 12 months of performance. The *Living Building Challenge* has a separate zero energy certification that has been added in cooperation with the New Buildings Institute. With a few exceptions, no combustion sources of energy are allowed.

None of these three useful programs directly addresses or includes facility cost in their rating systems, although the USGBC has provided some focus and research to study the cost of sustainable projects. An important study from 2007 by Morris and Matthiessen compared the project cost of LEED-certified projects with the cost of comparable noncertified projects for different building types.[6] LEED certification was used to indicate sustainability.

That work showed a large, almost identical range of cost per square foot for both certified and noncertified projects in different project types, but there was no appreciable cost difference between sustainable and other buildings. After adjusting the per square foot costs for year of construction and locational cost differences, the median costs for both groups of buildings were also very similar. The distributions were also similar. Two messages stood out:

1. You can build a very expensive building or an inexpensive building of either type.
2. You can choose to build a sustainable building at lower cost.

Despite this and other studies on cost, the myth continues for many that sustainable buildings are costlier on a square foot basis. This myth continues to hold back some building buyers from pursuing or asking at the beginning for a zero energy sustainable design.

The best antidote to the myth that solar sustainable schools must cost more will be more examples like the case studies that will be described later. In the meantime, the starting point is for a school district to plainly ask for what it wants—and not accept less.

NOTES

1. Thomas A. Taylor, *Guide to LEED® 2009: Estimating and Preconstruction Strategies*, Hoboken, NJ: John Wiley & Sons, 2011.

2. "Zero Energy Is an A+ for Education: Discovery Elementary," U.S. Department of Energy, Building Technologies Office, Zero Energy Case Study, August 2017.

3. ASHRAE, *Advanced Energy Design Guide for K–12 School Buildings—Achieving Zero Energy*, Atlanta: ASHRAE, 2018.

4. International Living Future Institute, *Living Building Challenge^SM 3.1: A Visionary Path to a Regenerative Future*, 2018.

5. U.S. Green Building Council, *LEED®: Reference Guide for Building Design and Construction*, 2013.

6. Peter Morris and Lisa Fay Matthiessen, *Cost of Green Revisited: Reexamining the Feasibility and Cost Impact of Sustainable Design in the Light of Increased Market Adoption*, Sacramento: Davis Langdon, July 2007.

Chapter 7

Encountering Fearful Buyers and Obstructionists

There comes the time in the planning for a new school, whether it be a sustainable solar school or any other school design for that matter, when the school district or private school leaders need to decide to go ahead with the project. This might be thought of as the point of no return.

Adventurous river trips have critical points. One of these points is when the parties joining the trip decide the river and date and have the permits if it's a regulated river like the Middle Fork of the Salmon River, the Selway River, the Colorado River through the Grand Canyon, and the Smith River. If you've agreed to the trip, you aren't going to let your friends down by backing out at the last moment.

These rivers also have another critical point: the point of no return once on the water. If you are launching at Boundary Creek, Idaho; Paradise Campground in Idaho west of Darby, Montana; Lee's Ferry, Arizona; or outside of White Sulphur Springs, Montana, there is no turning back for quite a while. You can't paddle upstream once you've started and you can't exit the river for some days because there is no access. The first place to reasonably hike out early on the Grand Canyon is at Phantom Ranch—after about seven days of paddling, if in rafts, kayaks, or dories. And there is no way to get the boats out of the canyon at Phantom Ranch. If you want to take the boats out, it's another week or more of river travel. The National Park Service, in any event, doesn't want your boat left there or anywhere else in the park.

School projects have similar critical points. The first point is when the school district has come up with a preliminary concept for the new school or major remodel and gotten the approval from the community to provide the funds, typically through a referendum. Passing referendums is a big deal. It's asking the community for a substantial financial commitment over many years. Many school projects are desired by school districts but never attempted as the prospects for referendum passage are too dismal. Others are attempted and fail; some fail repeatedly. Some pass.

The second critical point is the point of no return. It is the signing of the construction contracts when the construction drawings are completed and construction bids have been received, negotiated, and ready to be signed. At this point, the school district is committed to the project and hopefully is confident with the team that they have assembled.

School districts or private schools and their assembled project teams will frequently face a variety of obstacles and fears at these decision points, and even after they've embarked on the school project. It should be noted that these obstacles and fears are part of any school project, but the schools projects considered here are solar school projects.

An early obstacle will often be the persistent myth that green or sustainable buildings will cost more than a conventional design. Adding on-site solar can only make matters worse, or so it would seem. The fear of added cost can be an obstacle for both school boards and administrators and for building professionals on some project teams. Settling this question is unlikely to happen in a single conversation. A useful response is to go by example, especially examples that are like the solar school being considered. Such examples provide encouragement to move ahead and moving ahead brings many decisions.

Building projects have many aspects and an enormous number of decisions to consider. It's easy to be overwhelmed by the myriad of details over specific building attributes and their consistency with the community's goals for the project and ensuring sustainability in the design and construction. This complexity is evident in the USGBC's LEED guidelines for building design and construction. The current version, Version 4, includes almost 70 prerequisite and credit areas.[1] Although it is more narrowly focused on energy, ASHRAE's *Advanced Energy Design Guide for K–12 Schools: Achieving Zero Energy* is an equally daunting document.[2]

A school administrator looking at these guides for the first time might well be intimidated by the prospect of a zero energy sustainable school and might understandably jump to the conclusion that the cost of a sustainable school might be higher.

Reading about examples of zero energy solar schools at or below regional cost averages certainly will begin to allay the initial cost concerns. Far more compelling, however, are visits by school district personnel, boards, and project teams to schools that have achieved solar energy, sustainability, and cost goals.

The response to the cost concern is for the school district or private school, as it forms its project team prior to the start of planning and setting referendum amounts, to explicitly set its cost expectations at the conventional construction cost average for the region, without sacrificing the basic sustainability requirements described in the previous chapter. If a school district asks for additional higher-cost features not related to solar and sustainability—such as natatoriums, competition-level field houses, and high-end theaters—the cost expectations will need to be adjusted accordingly.

There is no doubt that some sustainable design features will carry a higher price tag. For example, geothermal (i.e., ground-sourced heat pump) systems, which are a leading option in general and particularly to avoid the use of fossil fuels in the form of natural gas, will typically add to the mechanical system cost because of the well field. There is also no doubt that other sustainable design features will carry a lower price tag in both first cost and ongoing operating cost.

An effective design approach for reducing system costs is the right sizing of the HVAC systems, especially in avoiding over-sizing the systems. Engineering designers are sometimes accused of using a belt and suspenders approach. They tend to leave a wide margin of error in their designs. Schools provide the opportunity to size mechanical systems recognizing a diversity of spaces that are not all occupied simultaneously. Providing system turn-down capability for partially occupied spaces is another important opportunity. Owners need to be informed of these opportunities and their ramifications. A robust, right-sized design carves cost out of the new school or remodeling project.

Reducing lighting power densities and fixture counts to what is required for lighting performance using Illuminating Engineering Society of North America (IESNA) and other professional guidelines has been a common cost-saving strategy in many school projects.[3] The evidence of excessive lighting in school projects can be seen in the emergence of professional services in lighting retrofits to reduce energy use and costs. These services came prior to the emergence of highly affordable, long-lived LED lighting, which has accelerated lighting retrofit work.

Another common cost-saving measure is related to glass selection in schools and other commercial buildings—to help manage glare and unwanted solar heat gain and to reduce the number and types of internal and external shading devices. Evidence of the opportunities in this area, where designs have not sufficiently controlled glare initially, is the professional retrofit services that have sprung up with window films to help manage glare and unwanted heat gain. The development of and growth of smart glass which changes light transmittance and solar heat gain to manage glare and unwanted heating provides another tool to reduce the need for and cost of window shading systems.

River Crest Elementary School in Hudson, Wisconsin, and Northland Pines High School in Eagle River, Wisconsin are examples of schools where glass selection was the primary approach used to manage glare, unwanted solar heat gains, and project cost, with the result that there is effective management of glare with very limited use of blinds or shades. This approach is in response to a common challenge in many existing schools where too much light is allowed through the glass during sunny and cloudy conditions. Light from windows is so bright that blinds are necessary. Blinds are shut, views are lost, and lights are now required as there is little useable natural light available unless the school has light tubes or other forms of top lighting.

With regard to on-site solar, the first cost burden is a challenge in that it is a building cost that heretofore has not been on the radar screen for most school planning. Most school designs were not even considering solar until recently. Thus, adding solar to a school project is an addition to the school project that school districts and their design teams had not been anticipating. While there are examples of schools that have included the cost of solar systems in their budgets, including some of the case studies considered in Section 3, this up-front cost can be avoided with the participation of third-party investment (TPI).

TPI enables not-for-profit school districts or private schools to leverage outside capital and monetize the federal tax credits and accelerated depreciation available to for-profit entities. The school does not need to obtain the up-front capital for the solar. Further, tax savings and accelerated depreciation reduces the solar project cost for the school. The third-party investors purchase and install the solar PV system and provide solar energy and sometimes other services to the school through PPAs (power purchase agreements), lease arrangements, or energy service agreements. Options to purchase the solar system in the future may be included.

The net result of the TPI is to provide electricity from solar without the first cost burden and at a lower cost than purchasing the power from the electric utility. Two-thirds of the commercial (i.e., nonresidential) solar capacity installed in the United States in 2016 involved third-party investors and that trend is anticipated to strengthen.[4] As will be described in an upcoming chapter, a variety of TPI options are available for schools in many states. As the solar market continues to mature, the trend will be for most, if not all, states to allow one or more TPI options.

The Northland Pines and Darlington Community School Districts have pioneered this approach to on-site solar in Wisconsin, which to date has been restrictive with respect to on-site solar. Wisconsin has low net metering limits and does not allow PPAs. In both cases, the school districts did not need to invest any of their own funds up-front. In the case of the Northland Pines School District, the cash flow was positive beginning in year one. While monthly payments are made on Northland Pines solar systems, these payments are less than the monthly savings from reduced monthly utility bills. The Darlington solar system was built a couple years earlier, when solar costs were somewhat higher. It is very close to cash-flow neutral and will become strongly cash flow positive when the school district buys out the solar PV system.

The goal in building a sustainable solar school is finding optimal mixes of design, materials, MEP systems, and solar PV that maximize sustainability while reducing or equaling overall first cost relative to conventional school construction costs. Paul Hawken, Amory Lovins, and L. Hunter Lovins have referred to this process as "Tunneling Through the Cost Barrier."[5] The financial bonus at the end of the process is a sustainable school with a first cost similar or lower that what it would have been with a conventional design and a considerably reduced energy operating cost and environmental footprint.

The evidence that this approach is working in a school project can be seen as the design unfolds and in the associated costs that emerge.

So, what are the fears and who are the obstructionists?

Fear is a natural and common part of design and construction. A primary fear is that construction costs will, in the end, be higher than estimated, and during the course of a project cost overruns will exceed budgeted costs. This is a problem for any commercial building construction project and is especially a concern with a fixed budget set by a school referendum. In a referendum situation, there is no choice but to stay within budget. In that respect, school projects might be somewhat easier to manage than some other commercial building projects in that the budget must be met.

If a sustainably designed school is being completed below the referendum amount, the school district will be looking to see what other things can be included in the project. An interesting example is the case of the Northland Pines High School that was coming in under budget, which already was well below regional cost averages. Some badly antiquated football field lights for night games were replaced as a last-minute addition to the project. The community endorsed the late change.

The difficult challenges and fearful aspects of school projects are finding solutions when project costs are not within budget.

Many design and construction companies that have transitioned to sustainable design can relate stories of fears and challenges in their early sustainable projects. An example is provided by Hoffman Planning, Design & Construction, Inc., of Appleton, Wisconsin, in one of its early, deliberately green, new high school projects in 2002. This was shortly after the LEED rating system had emerged and Hoffman was trying out the LEED guidance and process.

The new high school in Chilton, Wisconsin, incorporated an innovative, sustainable design approach guided in part by the work at the Energy Center of Wisconsin (now Slipstream). The LEED rating system had recently been introduced, and the project team used the guidelines to an extensive degree, although it did not attempt to formally certify the school. The lighting design engineer, however, was skeptical of the use of direct/indirect fixtures at greater spacing while meeting IESNA design guidelines but at markedly reduced lighting power density measured in watts per square foot. The design engineer initially refused the request for an alternate lighting design with the wider spacing.

Part of this fear was based on a substantial career experience of over-lighting space. The comment was that he had "never been sued for over-lighting a building." As the project neared completion, however, it was evident that the school was about $50,000 over budget. Focus returned to the lighting question, and the decision was made to reduce lighting power densities in all regular classrooms and a few other areas to IESNA levels to make up the $50,000. The lighting designer reluctantly agreed to use a lower lighting density design. Follow-up light measurements were made to verify lighting targets had been achieved. They had.

Another early green Hoffman project was for a biopharmaceutical client for which it had built about 30 clinics that were of identical design. An agreement had been reached with the client to deliberately modify the design to increase sustainability with better daylight and views, improved energy efficiency, enhanced indoor air quality, and other measures. Some of the project-team designers and construction managers were dubious that project costs of the greener design could match the cost of the previous design with identical square footage.

Thus, it was a bit of a surprise to some members of the project team, when the construction bids came in, to find that the project cost with improved sustainability had gone down by 2 percent relative to the previous clinic design. That resulted in a string of follow-up clinics in the more sustainable design.

A dramatic example of tunneling through the cost barrier in a sustainable school is provided by the 93,000 square foot River Crest Elementary School discussed in chapter 3.

As noted there, the average cost for conventional elementary schools in the region in 2008 was $223 per square foot. River Crest cost $129 per square foot. The total project cost including design was $166 per square foot. This was 26 percent below regional cost averages, even including the design costs. According to *School Planning & Management*, most of the schools in their database provide construction-only cost, excluding design.

While cost comparisons are notoriously difficult because few school designs are ever replicated, these and other examples from across the country of the financial competitiveness with sustainable projects are demonstrating what is possible with a disciplined project team and approach. A deliberate project approach can provide sustainability and financial benefit in commercial buildings, including schools.

These advantages can be obscured, at times, by counterexamples of sustainable projects with relatively high costs. As previously described, research on cost comparisons of sustainable or LEED certified to conventional buildings by Morris and Matthiessen shows that there is a wide distribution of costs for both conventional and sustainable projects.[6] The distribution and median values were nearly identical for the certified and noncertified. Building costs were controlled for both year of construction and geographic location.

The lesson from examples of highly cost-competitive, sustainable buildings and the research on building costs is that school districts and other buyers of sustainable buildings should recognize that cost is something that can be managed to the outcome they desire. Cost needs to be actively managed as part of the sustainable design and construction effort from the start of a new school project. At the very beginning of their project, school district administrators need to ask for what they want, including a desired cost target, and work with their team to get there. The cost level should never be left to chance or as the result of a design effort.

The joy of river trips includes the escape from regular routines including work and the daily accounting of time and costs. The focus becomes the river, the environment through which one is traveling. It's mesmerizing and reflective. The costs of the trip have been determined prior to embarking. If the costs were too high in funds or time, another trip would have been chosen or no trip taken at all.

Sustainable building projects can be great sources of satisfaction for school districts, their communities, the project team, and the numerous contractors involved. Even in these rewarding projects, there may be other obstructions and obstructionists to sustainable solar schools that are not related to building design and construction features and costs or to solar PV system costs. One example of obstructionists to new school or remodeling projects is a group of community members who believe that it's fiscally imprudent to undertake a school project where the existing school is argued to be sufficient, even when it's not.

It's no surprise that there is a diversity of views on school projects, starting with the big questions—such as whether the project is even needed, down to small points relating to specific project design decisions. Even in the case of school projects that are overwhelmingly needed and meritorious, many school districts will face a segment of the community that will present obstacles to a school project. This reality also applies in many other commercial building contexts.

A diversity of views can also come up within project teams and within school boards. Some types of obstructionists might be characterized as

- Technophobes and performance skeptics
- Absolutists
- Control freaks

People with these characteristics have been known to be encountered on river trips. The performance skeptic is that person who doesn't think the river can be safely run with certain other people on the trip or people with certain types of equipment, or at certain stream flow levels. The technophobe may be stressing about whether a new kayak or raft or dory can be safely used on the trip. These are fair questions, but sometimes they won't let the issue drop once it's been discussed and debated.

The somewhat opposite character is an enthusiastic supporter of some new boat design who is convinced that the technology is the only way to go and must be used. This character trait may also play out in other ways, such as how a certain rapid must be run, or the schedule and campsite selection on the river. The problem emerges when the absolutist won't let their favored technology, schedule, strategy, food menu, and other things drop if their proposal does not gain the acceptance of the group.

Control freaks or the *my way or the highway* types are difficult on group dynamics. Ultimately, the river travelers sort these issues out to a reasonable level before the trip, and/or deal with issues that still come up during the trip. No one would assume there would be no friction or issues on a river trip, but one would hope that these people issues remain within reasonable interpersonal relations.

School projects can also have obstructionists with these character types and others. The Northland Pines High School project was presented with alternate proposals for a high performance HVAC system. One was an established design and the other was an innovative design with proprietary agreements required. The innovate design system had been recently installed in one other school in the state. Data from that school revealed that the energy use in the innovative design was at least 10 percent above the model estimates for the established design. This difference between the two design approaches was later verified in actual operation of the school after construction.

The school district and their team ultimately selected the established design to meet their high performance requirements. The HVAC designer whose design was not chosen and others in the community were not willing to accept that decision. Not only did they believe theirs was the better system, but in their absolutist perspective it was the only HVAC system that could meet the school's goals. This conflict over the HVAC design lasted for months with multiple school board meetings and impeded the design schedule. The construction, however, was still completed on schedule.

Technology choice is complex. It encompasses everything from building shell features, especially window types and locations, to the design of the mechanical, electrical, and plumbing systems. The combination of building orientation, the locations of windows and skylights, and the choice of glass and electric lighting fixtures determine the daylighting and views that have been linked to student performance in schools and productivity in work environments. Debates over technology choice need to be limited in time as choices need to be made in a timely fashion to meet school opening deadlines.

Daylighting design has taken two broad approaches with strong advocates for both approaches. One approach advocates glass with high light transmission with the resulting contrasts managed with, external shading, window treatments, and skylight treatments. Another approach manages the glare with much lower light–transmitting glass, like what sun glasses provide, and less reliance on shading devices.

School buses have widely adopted the sun glass approach to controlling glare and unwanted heat gain. The approach to the management of daylight and glare is currently being further informed with the introduction of intelligent or smart glass that adjusts visual light transmission according to exterior light conditions. This is like photochromic lenses for eye glasses. As light levels and associated glare increase, the glass becomes darker and reduces light transmission. As with many new technologies, higher cost is the barrier.

The decision to include on-site solar PV systems and whether to use third-party investors can face obstructionists such as performance skeptics and technophobes. Skepticism and thoroughness in review by project team members and those outside the team are part and parcel of a rigorous school design and review process. Skeptics can become obstructionists, however, when they won't accept the validity of verified technical data on the performance of system components and the performance of existing systems in similar applications.

Performance data are available on solar panels and inverters using widely accepted standards. California is a leader in energy efficiency and renewable energy. A useful approach for solar PV system designs is to require panels and inverters that are on California's list of incentive-eligible PV panels and inverters.[7] Inverters are the electrical devices that convert DC (direct current) to AC (alternating current).

The rapid change in many aspects of sustainable school design, especially with solar systems, presents challenges for everyone. As these changes occur, designs quickly adapt. The change in the cost of solar PV equipment that has declined by about 75 percent from 2010 to 2018 is resulting in rapid increases in solar adoption. Changes continue to occur in financial incentives that vary by state and utility. Federal investment tax credits continue to change, and tax rules and rates for third-party investors change.

The result of this rapidly changing picture is that solar PV systems can provide positive financial returns that were unimaginable a few years ago. Some of the obstructionism may simply be that people are not keeping up with innovation and the market.

The irony of some of these obstructions is that disagreements over technology or design choices among members of the school project team or between the project team and community members or other outside parties may not even matter in some cases. There are often several design options that will support sustainable schools, but some individuals are so committed to the virtues of their approach, that they feel compelled to block alternative, viable approaches.

An example is a project design that was delayed for months by an individual on an outside expert project advisory committee who believed that what would ultimately become a LEED platinum project could not be sustainable given the external materials (mostly brick and glass, rather than wood and glass) chosen for the building shell and a design that was more rectangular rather than curvilinear. The owner ultimately decided not to go with a curvilinear wood and glass building. It was only many months later—after LEED platinum certification, energy performance as anticipated, and enthusiastic community response—that the advisory committee member embraced the new building.

Fear and obstruction are an inevitable part of school projects, and that applies to sustainable solar schools. There's not much choice but to anticipate that obstruction may come, persevere, and paddle ahead.

NOTES

1. U.S. Green Building Council, *LEED® Reference Guide for Building Design and Construction*, Washington, DC: U.S. Green Building Council, 2013.

2. ASHRAE, *Advanced Energy Design Guide for K–12 Schools: Achieving Zero Energy*. Atlanta: ASHRAE, 2018.

3. Illuminating Engineering Society, *The Lighting Handbook*, 10th edition, New York: Illuminating Engineering Society of North America, 2011.

4. Greentech Media Research, *U.S. Commercial Solar Development Landscape, 2016–2020*, May 2016.

5. Paul Hawken, Amory Lovins, and L. Hunter Lovins, *Natural Capitalism: The Next Industrial Revolution*. London: Earthscan, 1999.

6. Peter Morris and Lisa Fay Matthiessen. *Cost of Green Revisited: Reexamining the Feasibility and Cost Impact of Sustainable Design in the Light of Increased Market Adoption*, San Francisco: Davis Langdon, July 2007.

7. California Energy Commission, www.gosolarcalifornia.ca.gov.

Chapter 8

Solar PV Systems and Intelligent Schools

School energy efficiency has been the focus of energy management and sustainability for almost 50 years, since the 1970s energy crisis. Energy efficiency, if deliberately applied, can provide an energy use per square foot reduction in a new school by more than 50 percent since the designs of that time. Evidence of this efficiency potential can be seen in the equipment options, including LED lighting, energy-recovery ventilation, and increased efficiency ratings in electric motors that power pumps, fans, compressors, and other mechanical equipment. Other evidence is provided by the percent efficiency improvement that LEED Version 4 provides points for, which is up to 42 percent relative to a new performance baseline that is considerably above 1970 practice.

Despite that progress, the vision of a zero energy school was only a distant dream until the dramatic fall of solar costs since 2010. The distant vision has suddenly become reality.

What has burst on the scene are affordable, on-site solar PV systems. Solar PV systems consist of these few elements:

- PV panels: convert sunlight (photons) to direct current electricity
- Inverters: convert direct current to alternating current matching the power provided by the utility grid to the school
- Racking: provides mounting for panels on roofs, the ground, or parking structures
- Electrical and monitoring equipment: provide the electrical connection, control, and monitoring capability, including meters and safety shutoff

As many readers are knowledgeable of PV systems and information and images are readily available, further description is not provided here.

There are very few other practical, on-site renewable options, although wind in some locations might provide the basis for zero energy schools. There are schools with some wind power that can be found across the country, especially in locations with favorable wind conditions.[1] A very small proportion of school wind systems are larger. At 1000 kW, Spirit Lake Community School District in Iowa, for example, provides an early example of a larger wind installation.

To be the most cost competitive, wind turbines have continued to grow in size measured in kW or MW. The U.S. Department of Energy's Office of Energy Efficiency and Renewable Energy reports that the average size of new turbines for the wind power market was 2.3 MW in 2017.[2] This average machine is four to six times the size required for the average new school and would stand many hundreds of feet high. As such, they are located away from the immediate school grounds. The average installed cost is reported at $1,611 per kW for these very large wind machines.

At the scale of, say, 200–800 kW appropriate for most schools, however, solar PV costs are more than competitive with the installed cost for wind power. At this scale, wind costs are higher per kW than the cost reported above for large wind units. Solar's additional advantages include no moving parts (resulting in lower maintenance costs), less or no visual intrusion, no noise, easier siting issues with local authorities, easier integration with battery storage, and the ability to use roof space. These factors have made solar the preferred choice for on-site energy for most schools and many other commercial buildings.

To achieve zero energy, the solar PV system needs to provide an amount of power over the course of a year to meet the school's cumulative energy requirements for that year. There will be times when solar production exceeds power needs and other times when the school's power needs exceed the solar production—especially nighttime hours. The critical service provided by the electrical grid is to assure that a school's power requirement is met at any moment. For this reason, there will continue to be some utility bill even in the case of a zero energy school or an energy-positive school producing more power than is used.

An example of daily power use patterns is provided by Northland Pines High School. The data in figure 8.1 is from 2016, prior to a large solar PV installation in 2017. The upper profile in figure 8.1 describes the maximum power use in kW at a given time over the course of the year. The middle profile is the average power use at a given time over the course of the year. The lower profile is the lowest power use at a given time for the year.

These profiles are instructive as they describe when power needs to be provided and can be compared to the availability of solar power. The times of greatest power use at Northland Pines occurs during hot weather and school is in session. It reveals a steep climb in power use beginning at 6 a.m. The school is waking up, so to speak, with the school spaces being cooled,

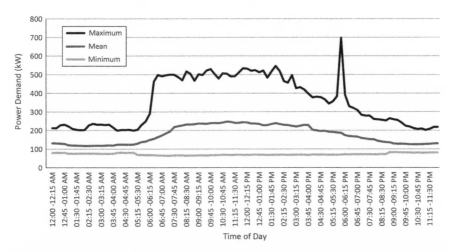

Figure 8.1 Northland Pines High School Daily Power Profile. *Source*: Data provided by Northland Pines School District: Graphic by Mark Hanson.

outside air supplied, and lights coming on. Power use then stabilizes through the course of the school day. Although ambient air temperatures climbed and internal heat loads including hundreds of students were added, the building power use stabilized until student activities fall off after about 2 p.m.

As will occasionally happen with mechanical systems, there was a problem with the chiller in the late afternoon on one of the hot days. The chiller was taken off-line briefly for repair. When it was turned back on, the chiller surged to full capacity as shown by the brief, but highest, peak demand. There was a large event in the school that evening and the school was getting back to desired temperatures prior to the event. Facilities staff later realized that allowing the chiller to do a fast recovery to return to desired indoor temperatures was an unnecessary operational error with financial implications, as there are demand charges associated with peak monthly power demand.

The mean power profile in figure 8.1 provides a sense of an average daily rise in power use that for most schools corresponds with the sun's availability or irradiance. Solar production begins shortly after sunrise, peaks at midday, and declines in the afternoon, ending with sunset. This makes solar power a particularly good match for schools as they have limited nighttime use. The amount of the daily rise in this mean daily profile is suppressed in a sense by the fact that Northland Pines High School has limited weekend use for community recreation and some athletic and theater events, and is only partially used in the summer. This reduces the average, daily peak-power use.

The minimum (i.e., lowest) power profile in figure 8.1 is likely from weekend days in the spring or fall with no requirement for chiller operation and perhaps no or limited heating. Power use is limited to ongoing ventilation, maintaining certain temperatures, emergency lighting, IT servers, and kitchen freezers and refrigerators, and so on.

As the size of a school's solar PV system increases, the amount of power exported to the grid or stored in on-site power storage (when available) grows. This is particularly true for weekends and holidays, when schools are mostly closed.

Having noted that a solar PV system for a zero energy school produces sufficient power over the course of a year to meet the energy needs, the question is this: Are there limits to how much power a school can import and export? If there is unlimited capacity to import and export power to and from a school, the only concerns in the school design are to provide a solar PV system sufficient in size to meet the aggregate energy requirement over the course of the year and estimate the financial results.

Public utilities are required to meet the needs of their customers. There is no limit on providing power to a school or to other customers. There are occasional limits, however, on how much power the school can sell back onto the grid. Prices paid for the power sold to the grid also vary, with the price often set at the avoided cost for the utility to produce the power, which are typically only the fuel costs at the power plant that is a small fraction of the cost of a kWh sold to the school. In other words, what price would the utility have had to pay to produce the power or to buy the next increment of power? The limits and terms for selling power to the grid vary from state to state and even from utility to utility within states.

The engineering challenge from the perspective of the utility grid in a region and for individual utilities is, How can they provide sufficient power at any moment? This is a real, long-standing issue. The regional electrical grid is continuously taking power from hundreds of power production sources, such as large, central, coal-fired power plants, natural gas–fired power plants, nuclear power plants, hydroelectric plants, and a few biomass plants. Since the 1990s, additional power has become available from new renewable sources—especially wind and solar. Wind and solar generation have made up more than half of all new electrical-generation capacity installed annually in the United States for all but one year since 2014.[3]

The addition of wind and solar power has added complexity in managing the grid in that the amount of wind and solar available is dependent on local weather conditions and time of day and can be variable over relatively short periods of time. Adding many solar and wind generators over a wide area helps to even out these renewable flows. Some solar installations can have bright sun while others are under clouds, reducing power production during daylit hours. Wind machines may be producing power at any time of the day, or not at all. The challenge is in balancing all the power sales and demands on the grid on a continuing basis. This is especially true with large baseload coal and nuclear plants in the power production mix that cannot be readily started and stopped.

As more, relatively small and highly distributed solar PV and wind generation is added to the grid from many households and commercial buildings such as schools, this source of power needs to be accounted for as well. The result of the cumulative effects of adding distributed and utility-scale solar generation has substantially altered California daily power pricing patterns on the grid. These patterns determine which power plants are brought on-line, starting with the lowest cost plants and moving upward.

Historically, hot sunny California days resulted in added air-conditioning load in the middle of day on into late afternoon. The power demand profile for the grid when it experienced air-conditioning load was somewhat like the school's highest day load in figure 8.1, except that the grid's profile would stay at higher levels longer into the afternoon than for this school. Power prices increased with these peak loads as peak power was the most costly to provide. This resulted in rising prices during the day with the highest prices in the mid- to later afternoon. After the peak loads were passed, the power prices would decline. This pricing reflected the increasing cost to provide power for peak needs for short-term periods.

With the growth of on-site and centralized utility solar capacity, solar is generally a good match for the air-conditioning load. Solar power production in California is suppressing power costs during the middle of the day. Late afternoon and after sunset present a challenge—especially in hot climates as power for air-conditioning continues but solar production drops off as sunset approaches. Many people return home for evening activities that add other demands for lighting, cooling, cooking, and various electronic devices such as flat screens. The result is that power prices now increase between 6 a.m. and 8 a.m., fall off during the day, and then increase between 5 p.m. and 9 p.m. The solution to economically providing some supplemental power before sunrise and after the sun sets appears to be energy storage, especially with batteries. These would augment and possibly supplant peak-load natural gas plants or hydro plants where available on the grid.

This issue of matching power demand and supply can't be avoided. With the growth of electric vehicles and batteries, energy storage is increasingly being considered as part of the solution along with changes in building design and operation to manage peak demands. Power can be stored in batteries from excess solar power generated during the day and from previously generated power—perhaps baseload power from the night before when power costs are low. The stored power can then be supplied to provide what otherwise would be high-cost power.

The response for schools and other commercial buildings includes adding intelligence to buildings along with energy storage such as batteries. The meaning of *intelligence* in this context is that the school has mechanical and electrical control systems such that the school can monitor its use of electricity and respond to the price of electricity on the grid in real time or in response to the school's fixed electric rate schedules. The response empowers schools to adjust operations to minimize energy demand during high-cost power times while continuing to provide thermal comfort.

River and lake wilderness travel have become intelligent since the 1990s with the development of communications and navigation aids such as global positioning systems (GPS), weather forecasting, and real-time data on stream flows. The purpose of added intelligence is to enhance safety and to access emergency services in the event of accidents. In the case of schools, the added intelligence is used to minimize energy costs.

Sea kayakers and canoeists touring the Apostle Islands of Lake Superior have added intelligence to remote travel with weather radios to plan safe crossing times between islands—especially where the crossings are exposed to the open lake waves from Gitche Gumee, or gichi-gami in Ojibwe.[4] In the event of fog or heavy rain, GPS is invaluable and eases navigation. More than one canoe party has experienced getting lost amid a maze of islands and bays in the Boundary Waters and Quetico on the Minnesota and Ontario boundary. While travel parties eventually figure it out with map and compass, handheld Garmins can now quickly take away the thrill of being lost.

Other types of intelligence in remote river travel are real-time data on river flows in parts of the United States.[5] River flow levels are useful data both in planning a trip as well as during a trip. The river flows will alter the hydraulics of a river. Some rapids will become easier or more difficult to navigate at different streamflow levels. At certain low or high levels, river travel becomes ill-advised and eventually impossible. Data on streamflows and long-range weather forecasts prior to departure provide guidance. Streamflow data can be accessed during travel where stream gauges provide upstream data points.

Certain rivers, including the Colorado through the Grand Canyon, have added releases at reasonably predictable times. This added intelligence enables river travelers to time the running of specific rapids at specific water flow levels. Trips with dories are likely to time their runs to specific flow levels at the Lava Falls Rapid. In cases where injuries occur in remote travel, another intelligent device is the satellite phone. Injured parties, for example, can be transported by helicopter from the Grand Canyon. Intelligent devices have increased safety and enabled rescue in remote travel. In the case of schools, the added intelligence continues to enable reduced energy cost and solar energy, no rescues required.

The electric grid has real-time pricing information, which regional system operators such as CAISO (California Independent System Operator) use to dispatch power generation. A few areas in the country have smart meters installed—that is, meters that provide the current price and even the anticipated price of power. Customers with building systems that can receive real-time pricing can then respond by reducing power demand and/or drawing on stored power.

Electric power price information for most schools is still limited to static tariffs at the present time. This is not a matter of technology as smart meters and intelligent controls systems and designs are available. The problem is that utilities are in the early process of providing smart meters in their distribution systems. Some utilities such as ComEd and Ameren provide smart meters to customers in Illinois.

What many electricity users such as schools are currently working with until real-time pricing is implemented in their area are fixed electric tariffs or rate schedules. While specific tariff structures vary considerably from utility to utility, they typically have higher on-peak pricing in cents per kWh in peak-use times (for example from 9 a.m. to 9 p.m.) and lower off-peak pricing for nights and weekends. Many commercial customers such as schools also have demand charges measured in dollars per kW over 15-minute intervals. The school's monthly demand charge is for the highest 15-minute interval during the month. The electric utility measures the kWh amounts and kW peaks each month and sends the bill.

It's common to have 30 percent to 50 percent of the monthly electric bill determined by the single highest monthly 15-minute kW interval. This pricing structure and how it can be used to manage the school's electric bill is often ignored or undervalued in school operations.

Some schools—especially sustainable schools—are getting smarter. An intelligent school will monitor kW in real time and respond to limit peak kW demands and monthly demand costs. The total school demand in kW in real time can be sent through a signal from the electric meter to the building control system (BCS). As the kW approaches predetermined monthly target levels, a few selected building operations can be modified on a short-term basis to limit demand. For example, temperature setbacks in selected areas of a school can be used to limit chiller demand for short periods of time of, say, two or three hours or fan power levels can sometimes be reduced. Some lighting might be reduced on a short-term basis, especially as excess lighting levels are common in many commercial buildings, including schools.

Intelligent options are also being added to schools regardless of tariff schedules. An example is light sensors to reduce electric lighting when daylight is sufficient and automatic lighting shutoff when no one is occupying a space. The same sensors can be used to adjust heating, cooling, and ventilation to non-occupied settings when no one is present.

The next step in school building intelligence is to take these capabilities that have been applied in response to fixed tariff schedules and augment them as utilities transition to real-time pricing. Demand management measures can be programmed to track electric pricing and batteries can be added.

Batteries provide the capability of storing excess power production from solar and using that power in the future as needed. This is especially advantageous for schools in areas where their excess solar power production would otherwise by purchased by the electric utility at a very low price, called avoided cost rates. For example, at one small school in Wisconsin, the price paid for renewable energy sold back onto the grid ranges from $0.038 to $0.056 per kWh, depending on if the use is on- or off-peak. If the power can be stored, it can be used later when the purchase price from the utility for that school is $0.134 per kWh, noting that in this case there was no separate demand charge.

In this school example, the serving utility has a net metering tariff. Net metering is a billing mechanism where the utility credits electricity sold to the grid at the same price as the customer pays for electricity. This allows for the school to have a solar system of up to 300 kW and net meter as long as monthly solar production does not exceed the school's monthly power use. Once that limit is passed, power is sold back at the lower rate. This school could get partway to zero energy within the net metering, but if it were to install enough solar to achieve zero energy, the operating costs would be too high because of the lower price to sell back a significant portion of the excess solar production. Moving any closer to zero energy would require batteries.

The use of stored energy extends beyond the use of that power just in the school. In a highly flexible, real-time, price-driven market that is only now being developed, it may include the value that the regional grid independent system operator (ISO) places on the stored energy for voltage support, frequency regulation, and related services. The electrical utility grid of the future evolves from just delivering power from generating plants to electricity users to also becoming an exchange service for buying and selling power and grid-support services.

How does a high-efficiency school become a zero energy school in a rapidly changing electric utility grid? There are four main principles to be followed:

1. Install as much solar PV capacity as is financially beneficial up to or beyond zero energy as interconnection and net metering rules allow.
2. Install on-site battery storage to maximal size as is financially beneficial, noting that a large battery financial benefits may come from grid support.
3. Plan to add additional solar, storage, and smart-building controls in the future as battery storage costs decline, electricity pricing and regulatory rules evolve, and the smart grid expands.

4. Use TPI to monetize federal investment tax credits and depreciation, manage operational risk, and eliminate first cost.

There is a wide range in the amount of energy required per square foot of space to operate a school on an annual basis and the amount is likely to continue to decline as efficiency improves for building shells, mechanical systems, and other loads in the building such as lighting and IT. Some of this range is due to differences in climate.

Schools in coastal California, Oregon, and Washington with limited need for heating and cooling will use less energy compared to schools in the upper Midwest and Northeast, with their significant heating needs in winter and cooling needs in summer. Schools in the South and Southeast have limited heating requirements, but substantial cooling requirements. Geographic locations requiring lower energy use will result in smaller PV systems and less roof areas required.

The developers of the ASHRAE *Advanced Energy Design Guide for K–12 School Buildings—Achieving Zero Energy* considered the roof area required for schools meeting the target EUIs in all ASHRAE climate zones.[6] The finding is that from 16 to 31 percent of roof area would be required, except for single-story primary schools in the Arctic zone where 45 percent was required. Even doubling the EUIs would still allow for sufficient roof area for zero energy. This suggests that many two-story schools will have enough roof area for solar, and schools also have options for solar over parking areas and ground-mounted solar where land is available.

High net metering limits allow some schools to proceed to zero energy without batteries. Batteries will become financially beneficial, however, in many locations with limited net metering. Battery and other power storage—whether it is by the utility or the customer—is just emerging. Thus, there will be a learning curve as this becomes available. The financially optimal battery storage capacity required to support a school will be heavily influenced by the utility pricing patterns for buying and selling power in different parts of the country and other rules, such as net metering limits within utility distribution territories and the revenue available for grid support services.

As a starting point, one might think in terms of providing an energy storage capacity for a school as equal to one day's use of electricity and for peak delivery capacity from the battery storage to approach the highest annual peak demand. The performance in the upper Midwest of a 100,000-square-foot elementary school with a geothermal system suggests a battery capacity of approximately 1,600 kWh and a power capacity of 400 kW.

Tesla's Powerpack battery was recently specified at 210 kWh (ac) of useful storage capacity with 50 kW of peak capacity.[7] Eight Powerpacks would provide 1,680 kWh of energy and 400 kW of power capacity. The cost of eight Powerpacks is volatile but approximately $1 million.

The overall first cost for the 100,000 square foot school is approximately $20–30 million in this example for the upper Midwest. The cost of the 600 kW (dc) solar PV system at $1,700/kW is approximately $1 million at current prices in projects in the Midwest. The cost of eight Tesla Powerpacks is approximately $1 million, but deliveries are just beginning, and the anticipation is that battery prices will decline, as has been experienced with solar panels over the last decade, as shown in figure 4.1. As battery prices fall, increasing numbers of these systems will provide financial benefits in schools that don't have high net metering thresholds.

While the additional $2 million cost for solar and batteries above the base cost of the sustainable school of, say, $25 million is not overwhelming, it's an 8 percent increase in cost that school districts are not used to budgeting. Third-party investors already provided the capital for well over half the solar capacity each year from 2011 through 2015. The TPI option in its various forms is available for schools in many states. This removes the first-cost burden of including solar and batteries in a school project. The goal now and moving into the future is to continue to identify the growing number of schools that can achieve a reduction in ongoing energy operating costs beginning in the first year of solar operation through TPI or acquiring solar energy.

The elements of a zero energy, solar-powered school exist in the building marketplace. The market for batteries and implementation of smart grids are emerging trends that are currently coming onto the market.

While all the elements of a zero energy, sustainable, solar school designed and built at regional cost averages exist, the optimal design with respect to MEP, solar PV systems, and batteries is and will remain in constant flux. The flux is the result of a set of factors that will continue to change including the following:

- Downward trends in solar PV system cost
- Downward trends in battery system costs
- Changing utility smart-grid infrastructure
- Changing utility tariffs and regulations, especially related to net metering and payment for grid services provided by batteries
- Changing carbon pricing and Solar Renewable Energy Credit markets

The optimal sustainable school design can only be set at a point in time at a specific location.

Hindsight from past sustainable schools has educated us as to the reality that making prudent, sustainable design and construction choices at any given point in time works out. Moving ahead provides desirable outcomes in terms of learning environments and financial performance. Other factors like student enrollment trends, facility obsolescence, and retirement of past school district debt usually dictate the timing for new schools or major remodels. Any consideration of delaying school projects for lower solar and battery costs usually results in increases in construction costs that are usually far more consequential to overall budgets than savings from waiting for solar or battery costs to decline.

Further reinforcement for moving ahead with solar as part of a sustainable school project is the ability to add additional solar PV and/or batteries in the future if conditions indicate phasing would be advantageous. This flexibility is provided with some planning. School roof design should anticipate future solar and avoid shading from adjacent building features and equipment. While self-ballasted solar PV systems are light—at usually less than five pounds per square foot—structural design should explicitly provide for this modest additional load. Some space needs to be reserved in electrical and mechanical rooms and outdoors for future inverters and batteries.

Solar is available for schools and should be planned for the present or near future. The next chapter considers the resources available to make solar energy a reality.

NOTES

1. Windexchange.energy.gov/windforschools.

2. U.S. DOE, *2017 Wind Technologies Market Report*, energy.gov, August 2018.

3. "Solar Energy Industries Association Solar Industry Research Data," SEIA.org, 2018.

4. Henry Wadsworth Longfellow, *The Song of Hiawatha*, Boston: Ticknor and Fields, 1855.

5. See waterdata.usgs.gov.

6. ASHRAE, *Advanced Energy Design Guide for K–12 School Buildings—Achieving Zero Energy*, Atlanta: ASHRAE, 2018.

7. www.tesla.com/powerpack.

Chapter 9

Technological and Financial Hurdles and Resources

There really aren't any technological or financial barriers that limit a river trip in the 21st century, even on more extreme and remote rivers such as the Colorado River through the Grand Canyon. There are a good number of kayak, dory, canoe, and raft options for self-propelled trips. The limited access and remoteness requires at least two weeks of time. Permits for self-guided trips can be hard to come by for some rivers, but that's not a technological problem.

There are new technologies that can be used on a river trip. As there is no cell phone communication, satellite phones can be brought for emergencies. Kayaks, paddles, clothes, and camping gear continue to evolve, bringing new capabilities, but older technology usually suffices. Water filters are a must and the options are many. There are costs, of course, in terms of permit costs, food, and equipment, but nothing extraordinary.

Similar conclusions regarding technology and financial barriers can be made for a sustainable solar school. We have the technology available now to build an intelligent, solar school. Newer and better technologies will inevitably emerge over time, but the tools available are sufficient. And the added first cost for on-site solar PV and battery storage can be avoided with the use of third-party investment.

A nagging challenge for many schools, including high performance schools, is the effective and skillful application of the HVAC, lighting, and other technologies built into the school. This challenge is perhaps like a kayaker on a river with class 4 and 5 rapids. The white-water scale commonly used spans from class 1 to 6. Class 1 rapids are very easy, with clear channels. Class 6 rapids are exceedingly difficult and risk the lives of highly experienced paddlers. The kayak, flotation devices, and paddle can be suitable, but is the kayaker sufficiently trained and experienced to competently and safety run and enjoy the rapids?

A challenge in many commercial buildings and schools, especially recent and generally more sustainable ones, is that the mechanical systems and controls are usually refined in design and competently installed, but the building controls programmers often fall short in understanding how the building is intended to operate and how to program the building controls to achieve that result. An additional challenge is matching the skills of building operators with those required to operate the building and maintain performance. Many buildings operate tolerably in terms of comfort, but often at an energy penalty in terms of extra cost.

A rule of thumb in the operation of existing commercial buildings is that a 20 percent cost savings would be achievable with the existing equipment or only modest changes if the controls were optimally programmed and mechanical components as basic as dampers were operating as intended.

These operational challenges are faced in schools and should be anticipated into the future as sustainable schools are designed with additional intelligence. The addition of more complex mechanical systems and controls can increase the challenge for the system designers, the programmers implementing the controls, and the building operators. The sweet spot for designers is providing enhanced control capability while keeping the control systems reasonably easy to use and maintain. Building operators require skill and training.

The addition of on-site solar does not present much of a challenge for building operators. The main task is to periodically review solar output to verify the system is remaining at full output consistent with ambient conditions. If the on-site solar is owned by a third-party investor, the owners will monitor the system. In either case, if production is reduced or stopped, the maintenance contractor will respond.

The addition of batteries might add complexity if the operators have choices to make in when to store, use, or sell power or make the battery available for grid series, such as frequency and voltage support. Third-party ownership will again transfer this responsibility to the investors.

The use of commissioning agents has become common in new school projects, additions, and remodels. Their purpose is to provide a process for designers and construction teams to evaluate, verify, and document their work to help assure that a school is designed to the owners needs and requirements, built as designed, and operated as designed.

A question often asked by school administrators is this: Why is a commissioning agent and their fee needed? Won't professional designers and operators provide a new school operating as we intended? It's a good question. The same thing might be asked of professional tour golfers. They're pros and certainly compete at the highest level. Yet even the biggest names on tour have their coaches. So should school projects. The commissioning agents also oversee the training by the project team of facilities staff in school building operation, including monitoring trends in important data that reveal how well the school is operating. Thus, commissioning has become a key element in surmounting the technological challenges in schools.

As described previously, on-site solar has been rapidly growing in terms of aggregate installed capacity (MW) and in the average size of solar installations at schools. In earlier years, solar systems were small in scale and costly on a per-kW basis. The rationale prior to about 2012 was focused on demonstration, education, and PR rather than as a cost-savings measure. As solar prices for panels as shown in figure 4.1 and entire system costs plummeted, the economics became advantageous and solar PV system sizes scaled up.

Solar procurement prior to 2012 used a mixture of funding approaches, including referendums, grants where available, and community fund-raising. Since then, third-party investment in its various forms has become a key resource and now is the most common approach to acquiring solar.

A key advantage of TPI is that investors can leverage or monetize federal investment tax credits and depreciation, which school districts and other not-for-profits, including private schools, cannot. Because part of the return to investors is provided in the form of tax credits, the net result for the school district is that it can achieve a greater reduction in electricity costs than with a direct purchase of a solar PV system. This is true regardless of which form of TPI is used, such as power purchase agreements, equipment leases, and energy services agreements.

The financial advantage is revealed in side-by-side comparisons of annual cash flows, NPVs (net present values), and IRRs (internal rates of return). Working with the solar providers or others on the project team, a school district conducts the necessary financial analysis and can compare direct ownership with its own funds or from borrowing versus TPI options available for the life cycle of the solar PV system. The analyses should be run for 25 years with NVPs also estimated at shorter time intervals such as 15 and 20 years, although the solar system may well be operated beyond 30 years. As long as the solar system is reducing the school's operating energy cost, there is no reason to stop operation.

Similar financial analysis should be run for on-site battery systems. Battery systems are eligible for investment tax credits when they are combined with solar systems. The expected life on batteries at this point is about ten years, thus the batteries should be analyzed bearing that in mind, with the presumption that the batteries would be replaced as required. As with the solar PV systems, the point is to reduce the school's operating cost whether the batteries are purchased by the school or TPI is used.

The arrangements for TPIs in states where they are allowed can take a variety of forms based on utility regulations in different states. One common form is a PPA (power purchase agreement). In a PPA, an outside entity owns and maintains the solar system at the school. An agreement is made for the sale of the solar power to the school at a set per-kWh price with allowances for inflation. This was the arrangement described previously for the 22 MW of solar systems at the Kern High School District in Bakersfield, California. The outside entity utilizes the tax credits, accelerated depreciation, and incentives (if any) to offer a long-term sale of power that reduces power costs for the school district and provides a financial rate of return for the investors.

Another form of TPI is a lease agreement. In a lease, the school district leases the on-site solar equipment and pays a fixed monthly lease fee. The investing entity that owns and maintains the solar equipment is paid a fixed monthly fee. The financial outcome for the school and investors is similar to the PPA. A lease agreement may be structured to include the transfer of ownership and maintenance to the school. At the point of transfer, payments to third-party investors would cease and all savings accrue to the school.

A third form of TPI is an energy services agreement. A fixed monthly payment is made for solar power and possibly related services such as demand management. An energy services agreement has been used in Wisconsin because of regulatory rules that appear to restrict the use of PPAs and leases. The desired financial result for the schools and investors is similar to the PPA and lease options. The energy services agreements have included buyout options beginning at year 12 or 15.

Purchasing the solar PV system from the third-party investors after, say, year 12, consistent with IRS rules, allows the school district to own and operate the system. From that time onward, the school district would no longer be making monthly payments to investors but would benefit from the solar production with the only cost being maintenance costs. If the school is not interested in the buyout, it can continue to make the agreed-upon monthly payments to the investors.

Northland Pines School District in far northern Wisconsin is an example of a school that uses an energy services agreement. The school district is certainly not located in what would be considered the Sunbelt. The schools are operated by an excellent facilities staff and are highly efficient, as indicated by their Energy Star ratings in the 80's and 90's. The high school built in 2006 was the first LEED-certified gold high school in the United States. The school district had a demonstration-sized 1.5 kW solar PV system installed when the high school was built.

The prospect in 2016 of a commercial scale on-site solar PV system was an exciting vision for the district, but the school board and administrator in this rural school district were clear up-front that they did not have cash reserves for investing in solar and that going to the tax payers for a referendum for bonds to purchase 418 kW of solar PV at three schools was a non-starter.

The school board was intrigued, however, at the prospect of a TPI-financed solar system and associated demand management measures to further reduce energy costs through an energy services agreement. The stipulation was that the solar project, demand management measures, and guidance on solar curriculum would require no school funds up-front and would be cash flow positive beginning year one. The positive cash flow is the result of the cost of solar power under the energy services agreement being less than the cost of the power that would otherwise come from the electric utility.

The administrator, school board, and teachers were particularly interested in the educational aspects of the project. The solar PV system and associated performance data provide a laboratory for STEM and other subjects such as business and finance.[1] With 260,000 jobs in the United States attributed to the rapidly growing solar industry, the solar PV system is a highly visible example for career opportunities in a broad range of areas from research and finance to engineering, design, manufacturing, and construction.[2]

As the financial arrangements for an energy services agreement were new for the Northland Pines School District and can be somewhat complex, the school district carefully reviewed the financial details, including a comparison of using third-party investors to a direct solar purchase. Specifically, they wanted to see how the monetization of the investment tax credits and accelerated depreciation schedules would benefit the school district. The benefit of the energy services agreement relative to a direct solar purchase was demonstrated by comparing the financial flows of the two options.

Figure 9.1 describes the annual and cumulative cash flows for a direct purchase of 330.4 kW of solar PV systems by the school district for the two schools that were at the same location in Eagle River. These are the Northland Pines High School / Middle School and the Eagle River Elementary School. After deducting a Wisconsin Focus on Energy Grant of $57,278, the school district would have had to use $454,057 of its own funds or borrowed the funds. The estimated cumulative cash flow after 25 years would be $970,000. The NPV (net present value of the system) at 25 years is estimated to be $301,000. These estimates are based on the assumption that the school district had the funds on hand, which as noted previously, it did not.

Figure 9.2 shows the cash flows with the energy services agreement. The annual cash flow is only slightly positive until year 12, which is the first option for Northland Pines School District to buy out the system. If it chooses to buy at that point at $124,600, the district owns the solar PV system and no longer pays the investors. The annual cash flow increases accordingly and by year 25 the cumulative cash flow is estimated to be $749,000. The NPV at 25 years is estimated at $284,000.

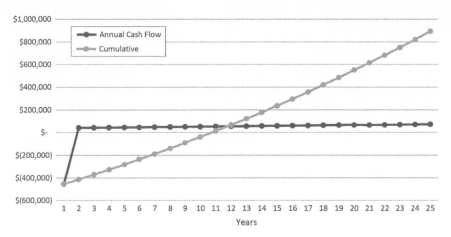

Figure 9.1   Northland Pines Cash Flow with Direct Solar Purchase. *Source*: Data provided by Northland Pines School District: Graphic by Mark Hanson.

The school district will need to plan for the purchase at fair market value in year 12. It can also defer the purchase to a later date. The cost of the buyout continues to decline with time, but the school only realizes a small annual cash flow as it continues to make payments.

The comparison of the cash flows between the two options demonstrates that accessing third-party investors who can use the tax credits and depreciation enables the school district to move ahead without having any funds to pay for the system initially. The cumulative cash flow and NPV are only

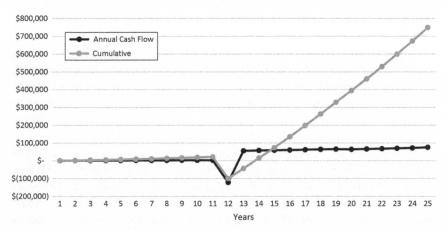

Figure 9.2   Northland Pines Cash Flow with Third Party Investment. *Source*: Data provided by Northland Pines School District: Graphic by Mark E. Hanson.

slightly below the case where the school had funds up-front. The school district's modified 25-year IRR (internal rate of return) was 12.2 percent using the energy services agreement, which is above the 7.2 percent modified IRR when using its own funds.

This financial analysis enabled the school district to move deliberately from initiating a solar planning effort in October 2016 to negotiation with third-party investors in the spring of 2017. After competitive bidding of the solar final design and installation, the PV system was installed and placed on line in November 2017.

This Wisconsin solar example encountered two other barriers to on-site solar that are common in some areas of the United States:

1. Seasonality in solar production
2. Net metering limitations in some states

The seasonality barrier is obvious in northern Wisconsin in winter with the combination of short days, low sun angle, and considerable snow cover. The 2017/2018 winter was particularly challenging with early and continuous snow cover with major snow events up to the middle of April.

Net metering is a measurement and payment practice where electric customers, such as a school that has on-site generation, can be compensated for the power that they produce. The glossary from the DSIRE database provides the following definition:

> For electric customers who generate their own electricity, net metering allows for the flow of electricity both to and from the customer—typically through a single bi-directional meter. When a customer's generation exceeds the customer's use, electricity from the meter flows back to the grid, offsetting electricity consumed by the customer at a different time during the same billing cycle.[3]

What makes net metering so helpful to zero energy schools is that they can size their solar PV systems to meet all their energy needs. When solar production is in excess of their needs, they can export the excess electricity to the grid and in effect use that exported power at some other time. When the net metering is at retail rates (the rate the school pays to buy power) and excess production carried forward to future months, the value of the exported power is the same as the power they purchase. This provides the highest value for solar production and the schools do not need to invest in on-site energy storage.

The use of net metering has spread widely in the United States starting with an early law in Minnesota in 1983. By 2018, all but a few states have mandatory net metering laws. Twenty states and the District of Columbia have net metering limits of 500 kW or larger. This includes the states of California,

Connecticut, Colorado, Delaware, Florida, Indiana, Iowa, Maryland, Massachusetts, Nevada, New Jersey, New Mexico, North Carolina, Ohio, Pennsylvania, Utah, and Virginia, and the District of Columbia. This means that schools in these states can have on-site generation capacity up to the limit in their state and be allowed to net meter. Many, but not all, of these states use full retail rates while some states use lower rates tied to the cost at which they generate or acquire power.

As previously noted, a large proportion of schools depending on their size and location would require from 200 to 800 kW of solar PV capacity. Thus, schools in some states are well positioned with respect to net metering to pursue solar for zero energy without requiring batteries or other storage. Even in the case of larger schools in states that are near the lower end of the range at 500 kW, they can at least get a long way toward zero energy without storage.

Wisconsin, in contrast, has lower net metering limits that vary from utility to utility. The net metering limit for the utility serving the two schools in Eagle River is only 20 kW at each school. With a buy-back rate of approximately 4.2 cents per kWh on-peak and 2.8 cents per kWh off-peak, the solar PV system capacity was limited by financial considerations to providing approximately 21 percent of the school's annual use. This amount of solar allowed the schools to avoid selling too much solar back on the grid at a loss. Providing 21 percent of electrical cost is a good start on solar, but still far short of net zero.

Substantial on-site battery storage as described at the end of chapter 8 would greatly assist in this situation in that it would enable schools to store excess solar on most days for about nine months of the year, even in northern Wisconsin. For the three months with limited solar production, there would be limited or no excess power to store on most days and power would need to be purchased from the grid. The availability of batteries, however, would enable the Northland Pines schools to purchase and store power at off-peak rates and use that power during the peak hours of operation and even sell power back to the grid on-peak or provide grid support services if that were made financially advantageous by the utility.

With net power purchases during three winter months, the schools would need to produce some excess power over the other nine months to achieve zero energy. What the next step might be on the path to zero energy at Northland Pines is described later in a case study.

At this point it should be evident that there are many different circumstances and a flexibility in how to approach solar schools. Some new and existing schools can choose sufficient solar for zero energy now and be financially ahead. Others will find financial advantage in going partway to zero energy. Adding additional solar, possibly with batteries, can be done in the future as conditions warrant. The approach in each situation is to optimize financial performance in terms of energy cost at a given point in time. Optimal solutions will continue to evolve with changes in the following:

- Electricity tariffs and the intelligence of the grid in communicating prices, especially for real-time pricing
- Net metering limits
- Installed solar PV system cost
- Installed battery costs
- Price paid for selling power onto the grid or for the value of grid support
- Availability of TPI options of various types
- Geographic region and associated climate conditions

One might protest that this sounds terribly complicated. School districts are used to figuring out their own best plan on many fronts. This includes everything from the design and remodeling of schools to serving the needs of a diverse student body.

It's useful to contemplate the polar ends of conditions for procuring a zero energy solar school in the case where the natural gas option is avoided so as not to emit any carbon. The Northland Pines School District comes close to the most challenging conditions. With a northern mid-continental climate, there is a need for considerable heating in winter when geothermal systems have their least advantage. There is also the expectation of providing air-conditioning, especially for summer school and community uses. The electric rates are about at the national average for commercial buildings, so solar and batteries must compete against utility-provided power in achieving cash flow neutral to positive outcomes using TPI.

The other extreme—where zero energy solar schools become a slam dunk—are places like the California coast with the combination of more temperate weather with minimal heating requirements and moderate cooling requirements. The electric rates tend to be far higher resulting in larger cash flows for solar and batteries.

Other areas that are comparatively advantageous for zero energy solar schools include the Northeast and South. The Northeast coast has appreciable heating loads although not as great as the upper Midwest, but has higher electric rates, which provides added competitive advantage for solar. The Southeast and South have modest heating needs, but greater cooling needs. Geothermal systems and even air-sourced heat pumps have the greatest advantage as mechanical systems in these areas. Their greater insolation nicely matches the greatest energy needs resulting in smaller solar requirements. Electric rates are generally lower, offsetting only some of the solar advantage.

It's no surprise then that the early zero energy solar schools and other zero energy commercial buildings are appearing in states like California, Florida, Kentucky, New York, Massachusetts, Maryland, New Hampshire, North Carolina, Oregon, Pennsylvania, Texas, and Virginia.[4]

Having considered the barriers to zero energy schools and some of the resources and tools available to overcome the barriers, the focus now turns to the solar school acquisition process.

NOTES

1. Northland Pines High School website link to solar monitoring.
2. SEIA 2017.
3. Database of State Incentives for Renewables & Efficiency (DSIRE), http://www.dsire.org/glossary.
4. *Getting to Zero Status Update and List of Zero Energy Projects*, New Buildings Institute, January 23, 2018.

Chapter 10

Acquiring a Solar School

There comes a go or no-go point in school planning and in river trip planning where the group is either going to push ahead or abandon the effort. The river trip idea can be revisited for a future year. The new school or major remodel can be reconsidered at some future date.

If the river trip is to go ahead for the coming season, the essential pieces are deciding who's going to participate and what river and specific section of river is going to be run. The group will have to decide on the river. This will be determined by many factors including skills, equipment, and past river travel perhaps with some of the same people. A critical piece of the timing will be the permitting requirements so that the group can secure the necessary permits and arrange time off from work.

As the trip elements come together, the group will have the conversation about whether this is going to happen or not. There may be a few members of the potential party that can't make it, but for those who make the trip, there comes the moment where the dreaming and planning becomes a commitment to go ahead.

Building a new school or a major remodel has parallel aspects but on a much grander scale. The time commitment on a new school can well be a four-to-five-year undertaking from the time of early needs assessment prior to referendum through to completion and monitoring of the early months of operation.

Some of the in-between milestones are

- Passing the referendum
- Designing—from schematic design through to final construction drawings
- Bidding of the construction elements
- Constructing the school, starting with site preparation and utilities, through to completion
- Commissioning completion and building turnover

Interspersed with these major milestones will be numerous design meetings, permit applications, construction meetings, and multiple phases of building commissioning. The reality is that taking on a new school project or major remodel is a massive, time-consuming endeavor. If and when the referendum has been passed, there is still two and a half to three years to get to the completed school.

Prior to a referendum, school district administrators and boards with a solar school as their goal will begin to assemble the early parts of a professional team that consciously brings on-site solar and sustainable design and construction expertise. The school district will do this on a competitive basis and evaluate the key capabilities and track records of architects, site planners, MEP engineers, construction managers, solar system designers and installers, and TPI providers. It is important to identify the capabilities of the professional service providers and verify alignment of vision.

The value of sustainable or green schools has gained increasing currency among school districts and administrators and in the design and construction community since the 1990s. As a result, it's common to see sustainability, high performance, green, or other similar phrases included in RFPs (request for proposals) and their evaluation criteria for selecting professional project services.

As financial considerations weigh heavily in referendums, school administrators and boards will collaborate with their nascent project team to develop concepts and associated budgets. Undertaking projects in the tens of millions of dollars, to as much as $100 million more in the case of larger schools, is a complex task. The concepts need to demonstrate how the proposed solar school concept meets the identified needs of the school district and provides a financially responsible path. Demonstrating that a sustainable, solar school is at equal or lower first cost than conventional design and that it will deliver lower operating costs can be powerful arguments for the community.

It is essential to start the process with a clear end point in mind. The endpoint is a net zero energy sustainable school using on-site solar PV systems. The cost target for total school project cost is to match or be less than conventional school cost in the region. In states where net metering thresholds are higher, this is already achievable using TPI. In states where net metering thresholds are low or have particularly low electric rates, the amount of solar may be limited by financial performance to somewhat less than zero energy. Battery storage capacity will increasingly be part of the TPI solar projects in areas with low net metering thresholds. The goal in new schools that are limited short of zero energy is to plan for additional solar and possibly batteries in the future as financial conditions warrant.

Some design and construction service providers, especially those that don't have previous experience meeting these requirements, may balk at this request. They may counsel and even lobby against such an ask at the outset of a new school project. One response to this pressure is the simple observation that if you don't ask for what you want, how will you ever get the outcome that you want? The school district or private school must begin the project with the end in mind.

The referendum process is critical to the endeavor. There can be no project launch without the referendum approval. As there can be no new school without identifying and acquiring a site, that may be part of the referendum process. Some level of concept development or preliminary design is needed to support the referendum, hence the forming of the pre-referendum team. For private schools, the concept development or design may precede the fundraising. The planning process includes identifying the school and community needs and preparing the case for the new school or remodeling. Gaining the endorsement from the community who are also the taxpayers is the essential go or no-go decision point. For private schools, it is common to attain a chosen threshold in fund-raising before proceeding.

School administrators are attuned to what referendum proposals are passing and which ones are not, especially in similar communities within their state. Many factors come into play in addition to fundamental needs for a school, such as the condition of existing facilities, school overcrowding, and life-safety concerns. Local area and regional economic conditions can heavily influence referendums. The 911 attack and its aftermath sank many referendums that had been planned for later in the fall of 2001 and on into the next year.

Referendum battles can be brutal and there are examples of school districts that seem unable to get over this hurdle. School districts face the referendum hurdle whether they are planning a school of conventional design, sustainable solar design, or something in between. From a strategic perspective, a school district or private school will need to come to the conclusion that a sustainable solar design provides educational and financial advantages. The school district would then make the case to the voters in the community or to donors at private schools. How the case is made and what points are emphasized will be adjusted for the particular interests and circumstances in the community.

If and when a referendum has been passed, the school district board and administrators will immediately turn to putting their long-sought plans into motion. This will include competitively soliciting and ultimately contracting with additional firms and individuals that will complete their project team.

While a clear, focused starting point is essential, so is the discipline during the project to adhere to this initial vision. Bait and switch is not acceptable. Most design and construction firms are well aware of the growing acceptance and downright popularity of sustainability. Their responses are highly likely to be positive. As the design process continues, there may be points of hesitation when one or more members of the team must face design approaches they may not have done before. It's at these points that the school district administrators and their teams need to hold steady to the original vision and not accept proposed changes unless consistent with and perhaps even an improvement on the end goal.

As noted previously, there are examples of solar zero energy schools that can be used to help persuade a skeptical service provider that zero energy sustainable schools are doable. There are additional examples of sustainable schools and other commercial buildings that meet the performance thresholds on the energy-use side but are only lacking the required solar capacity.

Drawing on examples of schools such as the Northland Pines and the Darlington School Districts that added substantial TPI solar well after schools were built as well as newer zero energy solar schools, school project teams should anticipate that on-site solar can be included in their project with no or modest up-front cost. The solar systems at these two districts leveraged about a 12 percent and 15 percent (of system cost) in state grants respectively to enable the school districts to install and co-own solar PV systems without initially committing any of their own funds. The ability to access TPI solar is restricted in some locations, thus the team will need to complete its due diligence.

Schools in other states may not have similar grant programs available, but many states provide other TPI options, such as power purchase agreements that don't involve ownership but result in operational savings. Another option may be to lease on-site solar PV systems. As long the lease payments and residual utility payments for grid power are less than purchasing all the required power from the utility, the school district will be financially ahead with PPAs and leases.

States and regions tend to have somewhat different patterns with respect to design and construction contracting arrangements. There are also differences with respect for requirements for using union shops. Some states or school districts will require union labor while others may use open bidding, where union and nonunion firms may bid. Sustainable projects are achievable with any of these contracting arrangements and requirements for use of union labor.

The primary distinction in contracting arrangements are approaches that separate the design and construction services into two separate and distinct activities versus contracting arrangements that deliberately seek to link the design and construction elements. The former is called *design-bid-build*. The latter is often referred to as *integrated project delivery*.

Under the design-bid-build approach, the school district selects an architecture firm that leads the design team. Once the design is complete, the school district provides the design documents for competitive bidding for construction. A general contractor or a construction manager is contracted for the construction. The challenge in a design-bid-build approach is that the design and construction firms have no relationship during the planning and design as the general contractor or construction manager become involved much later in the project. These team members can end up being adversaries rather than a team with a common goal.[1] Finger pointing can ensue with designers blaming the construction teams for issues that emerge and/or the construction team blaming the designers for design inadequacies.

Many involved in sustainable projects including the USGBC have evolved a strong preference for integrated project delivery arrangements. The purpose of integrated project delivery (IPD) is to provide a team approach to design and construction services rather than having different firms under totally distinct contracts. Various approaches are utilized in IPD, including the use of multiparty agreements, temporary organizations, and permanent firms that offer combinations of design and construction management (CM) at risk where all the construction elements are competitively bid.

The full design and construction process commences with bonding and contracting. From that point, the design process requires 9–12 months and construction typically 18–24 months.

Regardless of the specific contractual arrangements, deliberate effort is required to engage both design and construction professionals at an early point to gain their knowledge and guidance on design, engineering, materials, constructability, and cost. The construction professionals provide early cost estimating and ongoing collaboration with design professionals on strategies to improve design and reduce cost.

Sustainable school designs compared to more conventional designs will have some building elements that will cost more, but others that will cost less. The school district or private school administrators are at the center of this discussion as design and construction professionals work together to identify and explain choices and trade-offs. The administrators, along with members of the project team, are taking ownership of choices.

An example of how this works was a school project where the budget was pushing up against the referendum threshold. As the project was being fine-tuned as the early construction proceeded, the option to reduce chiller size by 15 percent was suggested. The worst risk identified by the mechanical engineer was that the temperature in the large field house might not be able to hold 74°F on graduation day with 3000 people *if temperature and humidity were at or above design conditions (worst case scenario)*. The consequence would be that by the end of graduation, the indoor temperature might increase a couple of degrees. The district administrator thought about the risks and trade-offs, and decided for the reduced chiller size.

After eleven years, the worst-case situation has never occurred.

The design and construction process is a long journey, with many decisions to make. While the launch of a school project is passing the referendum, the end of construction should not be thought of as the take-out point. The take-out point occurs after operations have been monitored and fine-tuned over the course of a year of operation. In the space between the launch and the takeout, there are many rapids to be run. Some need scouting where boats are pulled to shore, and the team walks downstream to check out the rapids and the best path through them. Others can be read as you go—especially if there is one or more people who have run this river and can serve as guides.

Essential guides in the process of acquiring a solar school are the energy modeler and the commissioning agent. The energy modeler uses simulation models to understand how the new school or remodel will perform and evaluates design options for the school administrators and their team regarding the building shell, MEP system components, solar PV system, and the controls for an intelligent school. In the end, the modeler and commissioning agent provide information on how the school will operate and estimate how much energy the school with its occupants and all their equipment will need over the course of the year. The project team, especially designers, can respond to the model output with alternatives for the school administrator to consider. When the design is finally settled, the estimated energy requirement then determines the capacity of the on-site solar and batteries if required.

The commissioning agent provides and oversees a process for the design and construction team members to evaluate, verify, and document their work in meeting the owner's building objectives. This results in a sequence of services over time:

- Assist the school district in laying out in writing the requirements (this is called *owner's project requirements* or OPR) for the school in terms of performance, especially as relates to MEP systems.
- Verify the design is consistent with the OPR.
- Verify the equipment installed is consistent with the design.
- Verify the start-up and operation of the equipment.
- Verify the school is operating per the design.
- Review performance during the first year of operation to verify it's operating as intended or explain why not and whether that should be acceptable to the school district or should be corrected.

This is a critical set of tasks.

Many building buyers ask why a commissioning agent is needed. Isn't the design and construction team a professional group fully capable of delivering a properly operating building without a commissioning agent? The answer in theory is, or should be, yes. They are professionals, but to enable the best performance from the team they need a highly skilled independent and objective reviewer to be watching the complex process of school design and construction. For golf fans, the analogous question might be this: Does Rory McIlroy, Brooks Koepka, Ricky Fowler, or Phil Mickelson need a coach? Apparently, they do, for they always have them.

Part of the rationale is the different orientation and detailed knowledge commissioning agents provide. They think through the entire building with its many systems and how they are intended to operate and interact. Given the nature of their work, they see many more and different buildings than architects and engineers will typically see. In some respects, a highly experienced commissioning agent will have seen it all. From a financial perspective, that extra amount of time spent on details by the design and construction team members takes time, and by implication, reduces profit. The commissioning agent isn't saddled with that internal conflict of interest.

Commissioning agents will monitor the operation and performance of the school in the early months of operation. The general term for this is *monitoring and verification* (M&V), which has become an increasing area of focus in buildings, especially with growing interest in green or sustainable buildings. The design team will have developed the M&V capability in the building control system as part of the design process in conjunction with the school administrator, facilities director, and others at the school district. Decisions will have been made on what elements of building performance are to be monitored and how.

The commissioning agent provides review and comment for the school administrator and staff. The agent will watch the sustainable school operation in the early weeks and months. Given the dramatically different conditions the school must operate under during the year, more so in some parts of the country than others, the transitions from heating season to cooling season and vice versa will be an area of focus.

At the end of the first year, the performance should be reviewed and compared to expectations. And the evidence will be there for whether the solar school met or surpassed zero energy performance in the first year. These reviews are not matters of energy use, energy bills, and carbon footprints only; they obviously extend to questions of comfort and the ability of the different spaces within the school to serve their purposes. Occupant comfort surveys can be a useful tool in this process. After the first year of operation, the monitoring transitions completely to the building operators, although some design firms monitor the performance of their projects over many years as part of a continuing improvement process.

Another important topic in acquiring a solar school is the question of third-party certification, such as LEED, Green Globes, and the Living Building Challenge. The question is often asked whether the extra effort and cost of certification is worth it. Mary Bowen-Eggebraaten, the District Superintendent for the Hudson School District in Wisconsin when the River Crest

Elementary School was built, had an elegant response to this question while reflecting on her district's decision to LEED certify. *Because we thought differently (after deciding to certify), we acted differently.*

Having considered the acquisition process for solar schools, we'll next consider case studies of five solar schools. These case study schools are a mix of zero energy–capable schools with substantial solar and zero energy schools.

NOTE

1. Thomas A. Taylor, *Guide to LEED® 2009: Estimating and Preconstruction Strategies*, Hoboken, NJ: John Wiley & Sons, 2011.

Section 3

FINDING THE MAIN CURRENT—CASE STUDIES

Chapter 11

Northern High School

Northland Pines High and Middle School is an imposing, brick and glass building that seems to emerge out of nowhere in the Northwoods. One would have driven at least 20 miles through white pine and hardwood forests, and tamarack bogs, to reach the town of Eagle River, Wisconsin. The high and middle school, along with the adjacent elementary school, are at the northern edge of town, tucked behind the snowmobile track that is the host of the World Championship Snowmobile Derby.

The first impression is that this might be one of the less likely places for a sustainable solar school. As noted previously, this is a more challenging climate than most to locate a zero energy school with an ASHRAE climate zone of 6a and annual heating degree days averaging more than 9000 using a 65°F base temperature.

Adding to the challenge for a solar school location is the previously described snowfall that, combined with long stretches of continuous temperatures well below freezing, means that solar panels can be covered with snow for weeks or more. Lake effect conditions from Lake Superior, fewer than 100 miles to the north, result in additional cloud cover during winter. To add insult to injury, the upper Midwest gets hot and humid in summer, which calls for cooling for most new schools, especially ones with summer use.

The argument for why the Northland Pines High and Middle School make an exceptional location for considering solar schools hinges on these conditions as well as the reality that this is not a wealthy community. If solar schools can be demonstrated at a school in this location, it can be argued it will be achievable in most other geographic locations that face less challenging climatic conditions and perhaps fewer financial constraints.

The fact of the matter is that Northland Pines High School and Middle School is not yet a zero energy school. In 2017, 230 kW of solar was added but Wisconsin's low threshold for net metering blocked going to zero energy on a financially advantageous basis for the time being

What Northland Pines represents, however, is a situation that will be representative of many existing schools in the United States. It's a school that is well administered, is widely supported in a fiscally conservative community, and well maintained. It's a school that will continue to work in deliberate steps to reduce annual energy costs and would like to get to zero energy. There's a reasonable chance it will get to zero energy by the year 2030, the year the 2030 Challenge targets for new buildings. If not by then, zero energy will likely come later.

The beginning point of this case study is a failing high school building, ill-suited for the Eagle River climate. Despite the sustainable virtues of reusing buildings, this was not one of those cases and the best that could be done was reusing materials and eventually diverting more than 95 percent of the materials from this building from landfill.

By the early 2000s, District Administrator Dr. Mike Richie and the Northland Pines School District Board had gotten to the point that they realized that a new high school was badly needed. They wanted a new school that would provide the types of learning spaces the district needed. They wanted a school that would provide much better aesthetics and indoor conditions, including daylight and views, air quality, and acoustic performance. It would be energy and water efficient. Middle school students would be added nine years after completion to create the now combined high and middle school.

The citizens of the area prize the environment and many are conservationists. It's a matter of culture, in part, but also reflects lessons of when the tall timber of the region was cutover between 1880 and the Great Depression, and the value of preserving forests, streams, and hundreds of lakes for pleasure and recreation—including fishing and hunting—grew. The economy is now highly dependent on tourism and vacation homes, and protecting the environment is protecting local natural assets that support a lifestyle and tourism. The citizens like the Northwoods, except for bouts of cabin fever come late February.

The school district used a competitive solicitation process to begin forming its project team to design and build a new high school prior to a referendum. The school district sought firms that were experienced in green design and were open to the concept of integrated project delivery. After interviews and meetings, it selected a firm, the Hoffman Corporation of Appleton, Wisconsin, for preliminary design and the referendum preparation work and potentially the follow-on project.[1] Hoffman had planning, design, and construction management professionals and was experienced in integrated project delivery.

The referendum goal was to provide the voters in the 474-square-mile school district with information that would enable them to support the deconstruction of the failing old school and build a new sustainable high school. At the end of the planning effort the decision was made to go forward with a referendum for $29 million.

Following the successful referendum, the school district board decided to continue with Hoffman for the full design and construction management. The project team utilized an integrated delivery approach with both design and construction professionals working together from the early stages. This in part is why the project team was selected for the pre-referendum work. The construction management approach competitively bid all the construction work to help achieve cost competitiveness.

Consistent with the region's values, the school district wanted a school with functional and even inspiring learning spaces. It would also serve the community with recreational, artistic, and other community activities. The school would be energy efficient and built with materials and systems that minimized maintenance costs. Sustainability was a concept that the school district embraced. The LEED program was gaining momentum in the commercial building market when the high school was in the planning phase and the LEED Guidelines were used as a guide beginning in early design. After much deliberation, the district decided to seek LEED certification and worked with the project team in setting a goal of Silver under LEED Version 2: New Construction.

The Hoffman project team, led by architects Tom Cox and Jody Andres made a commitment to the school district that, with the help of some incentives from the Wisconsin Focus on Energy Program and other sources, LEED certification, commissioning, and a very small solar system would be funded within the $29 million referendum amount.

The 250,000-square-foot high school and de facto community center was completed in August 2006 on budget. The project team recognized in the late design and early construction phases that a LEED Gold certification might be possible. A combination of some insights from the project team, including essential information from the energy modeler and commissioning agent, and discipline in decision-making by the district staff and school board, supported a push to certify at the gold level. In the end, Northland Pines became the first LEED Gold public high school in the United States.[2]

It's hard to imagine that any large, complex project like a new school will not have challenging moments. Some of these issues are related to the design and construction activities, while others are what might be called people issues. Various forms of interpersonal conflict arise out of different understandings and values. These challenging moments are the rapids, snags, or sweepers encountered following the project launch.

One small snag in the project occurred when a painting contractor substituted a paint on some of the walls that did not meet the LEED VOC (volatile organic compound) requirements. The error was caught by a project team member early in the painting effort, but just late enough that even with the application of the incorrect paint stopped, the average VOC levels from all the paints and finishes in combination were just over the limit for one of the LEED credits. One credit was lost as the project was trying to close in on a gold certification.

More serious rapids were encountered. One occurred when a mechanical engineer, who had not been selected in the early competition to be part of the project team, challenged the selection and design approach of the engineer chosen for the work. This turned out to be an extremely long rapid in that the episode took place over many months, starting with an appeal to the school board. When that challenge failed in changing the school board's mind, the engineer who had not been selected and a few individuals from the community went to court contending that the design that had been submitted did not comply with state code requirements and thus should not have been issued a building permit. This challenge was denied.

School project budgets are an ongoing challenge. Once the referendum amount has been set, there is no choice but to meet the budget. This project was particularly challenging in that its aggressive budget at $29 million for a 250,000-square-foot school was well below regional cost averages. In the end, Northland Pines High School was designed and built for $116 per square foot, including site work and deconstruction of the old school. This was 25 percent below the regional cost average of $154 per square foot.[3]

The challenge in these situations is finding cost savings without sacrificing functionality or compromising the desired LEED rating. The school district and its project team made some astute moves as they navigated the rapids. One example was the shift from a conventional brick to a less costly concrete brick. This shift had sustainability advantages, including local sourcing from a nearby manufacturing plant and a greater durability.

Another example of cost savings was the careful evaluation of chiller sizing. After engineering analysis and modeling, the project team estimated that the chiller size could be reduced. Engineers tend to like having an extra margin of safety in sizing equipment. The judgement becomes, When does the margin become too much? This led to a conversation with the school district administrator Mike Richie over the question of what the risks would be if a smaller chiller should be slightly short of capacity at extremely hot and humid weather conditions.

The worst-case scenario would be an early June afternoon at design weather conditions or worse for Eagle River with a high school graduation in the field house with 3,000 students and families present. Wisconsin Code design conditions are 86°F dry bulb and 75°F wet bulb. The field house would be set at 74°F, and if design conditions were anticipated on a graduation day, the field house could be pre-cooled to, say, 72°F or 73°F. The risk would be that the field house might not be able to hold at 74°F but might drift upward by a degree or two during the graduation ceremony.

Mike Richie's response was to immediately accept the risk and reduce the chiller size for project cost savings. It was sort of a practical, *get real* response. That slight probability of some discomfort simply wasn't a risk worth sweating, especially in a state where half the schools did not even have air-conditioning. After eleven years of operation, the chillers have only been run to capacity on one occasion and that was in error after an outage had temporarily shut down the chiller and the chiller was being restarted.[4]

Northland Pines High School has lived up to the expectations since completion of the project in 2006. Energy use and costs have met expectations with overall electricity use slightly lower than modeling results while natural gas use, which accounts for about 30 percent of costs, has been slightly higher. The Source EUI is 117 kBtu/ft^2 per year. This energy intensity is comparable, despite the climate disadvantages and summer use, to schools and other buildings in the NBI tabulation of zero energy verified and emerging schools.

The missing element in achieving zero energy, however, was the lack of on-site solar. A 1.5 kW solar PV system had been included in the original project as a demonstration and education system. That was as much as was feasible given a constrained budget and the costs of solar at that time, which exceeded $8.00 per watt of capacity. The hope was that at some point in the future solar would be affordable enough to compete with utility power.

That day arrived sooner than anticipated and with a financial arrangement that had not been anticipated in 2006. By late 2016, solar PV system costs had declined to under $2.00 per watt for systems in the range of 200 to 800 kW in Wisconsin. As solar prices had rapidly declined, investors were entering the solar market in substantial numbers. Inevitably, investors developed business models to utilize federal investment tax credits that were not available to nonprofits such as public and most private schools.

This was also true in Wisconsin, despite a restrictive utility regulatory environment. One straightforward path for school districts in many states is to find investors who would invest in the solar system and use a PPA (power purchase agreement) to sell the solar power to the school. In states such as California, Illinois, and New York this is done. In Wisconsin, however, the Public Service Commission of Wisconsin has deferred on whether a PPA is permissible and decided that allowing PPAs would require action by Wisconsin Legislature.[5]

Investors, however, began alternate approaches using energy service agreements for solar and complementary demand management services. The specific approach presented to Mike Richie and the school district's finance committee had been used in previous Wisconsin projects, including the Darlington Community School District, the Madison Country Day School, which was a private school, and Holy Wisdom Monastery in Middleton. Mike Richie and the school district were convinced by the successful precedents and decided to take a serious look at solar. By December 2016, the school district decided to move ahead with a phased planning study working with Hoffman, the lead firm that had worked with the district on the new high school, and Madison Solar Consulting.

Phase 1 of the planning study would be a feasibility analysis and a grant application with the Wisconsin Focus on Energy Program. A key element in this process is that the school district asked up front for what they needed to make the project feasible. Those three conditions were the following:

1. The solar projects would need to be built with no district funds.
2. Solar projects would be evaluated at the high school, which had just absorbed the middle school, as well as the three existing elementary schools in the district.
3. The solar projects in aggregate would have to be cash flow positive beginning in year one.

If those steps proved fruitful, Phase 2 would be the solicitation for final solar design and installation, and identification of investors willing to finance the project. If Phase 2 worked out, third-party investors would be identified and the contract agreements for installation would be finalized and executed, and interconnection agreements set up with the serving utility companies.

Before proceeding on any stretch of river with substantial hazards—such as rapids; flooding potential, say, in the case of convectional storms and unfavorable topography; extreme isolation such that rescue opportunities are limited or not readily available—river runners assess the various risks. The various parties to this solar project went through their assessments, and all the parties decided to go ahead at two of the schools.

In the case of the school district, a financial risk occurred during the early planning phases. If the solar project proceeded, the planning costs would be folded into the TPI financing arrangement. If, however, the school district decided to stop the solar planning effort at the end of phase 1 or phase 2, the district would be on the hook for those modest costs. The main concern would be that the projects would cash flow in the negative or be underwater as both accountants and river travelers would say.

Mike Richie and the school board decided to take that risk, based in part on detailed conversations, the precedent of other solar projects done in this manner in Wisconsin, and a high level of trust in the solar project team that carried over from the 2006 new high school project. The results of the phase 1 work done in the winter and early spring of 2017 strongly indicated that the three conditions in the Northland Pines School District ask could be met with one exception.

The exception was that one of the elementary schools was not very suitable for solar due to site conditions and the utility rate structure at that site. The site had tree shading issues and was a small site. A solar project there could not achieve cash flow positive in year one, and indeed would be cash flow negative for the district until year 12 when the school district would have the first option to buy the system from investors. If the district purchased the system in year 12, cumulative cash flow would eventually go positive. While the district was disappointed that not all its schools would be solar, they understood the underlying constraints.

Phase 2 presented two additional project risks. One was the matter of final solar installation bids. Phase 1 estimates were based on preliminary bids by two solar firms known to the project team from previous solar projects. Phase 2 bids, however would be the actual bids, based on the latest market conditions. Phase 2 would also provide the final terms for the third-party financing, including cost of the investor's bank loan as part of the financial package.

In a way, it's like running a rapid on a river. What the guide books say and what even the experience was on a previous trip by the same group will change with the actual conditions such as water flows or recent flood events that may have altered the outwash from a side canyon.

The final bids for the solar system and the financing terms came in slightly better than expectations and enabled the school district to proceed with the solar projects. The aggregate size of the solar systems at three schools is 418 kW. 330 kW of solar is operating at two of the schools. A different serving utility at the third school at Land O' Lakes has proposed that it be the third-party owner of the solar system on favorable terms for the school district. That installation is pending the results of that negotiation.

Figure 9.2 shows the annual and cumulative cash flow for the solar projects in combination at the two Eagle River school sites. The cash flow is for the case that the school district selects to buy out the solar system at the first option in year 12. As shown in the figure, the school district will have to plan to set aside some cash in year 12. At that point, the ownership and O&M will transfer from investors to the school district. The cumulative cash flow is estimated to be more than $700,000 by the 25th year, the year that the solar panel warranties end.

These solar systems are anticipated to be operating for 30 to 40 years. The financial estimates include O&M costs that anticipate inverter replacement at some point not too far after their warranties expire in year 20. Should these expectations be meet, the cumulative cash flow could approach $1.5 million by year 35.

The Northland Pines School District is anticipating another round of solar investment in 5–10 years, driven by the development of battery technology to store energy on-site. The ability to produce excess power and store it for future use and sell excess power and grid support services to MISO (Midcontinent Independent System Operator) will make that a positive financial choice for the district. It's too early to know if the next round of investment will fully move to zero energy or will be another intermediate step on the path.

These three schools ranging in age from 12 to 20 years have the roof area and/or land to get to zero energy. What is holding them back at the moment is further development of smart meters and reduction in battery costs. An increase in Wisconsin's net metering threshold to, say, 1000 kW would immediately alter the situation.

Northland Pines demonstrates how even existing schools in a location with challenging climatic conditions and reduced insolation can utilize solar to move toward zero energy in a manner that does not require up-front capital and is financially beneficial.

NOTES

1. Hoffman Corporation was succeeded by Hoffman Planning, Design & Construction, Inc. in 2012. See www.hoffman.net.

2. Mark E. Hanson, "Green Schools on an Ordinary Budget: The Cost Case of Eco-Friendly Design at Two Wisconsin School Projects," *The School Administrator*, August 2010.

3. *12th Annual School Construction Report*, *School Planning & Management*, February 2007.

4. Interview with Dave Bohnen, Facilities Director at Northland Pines School District, 2017.

5. Letter from Robert D. Norcross, Administrator, Gas and Energy Division, Public Service Commission of Wisconsin, to Representative Gary Tauchen, February 8, 2012.

Chapter 12

East Coast Elementary School

Discovery Elementary School in Arlington, Virginia, has received considerable, well-deserved recognition as a zero energy sustainable school. The
awards include the 2018 ASHRAE National Technology Award and the 2017
AIA Cote Award.[1] Discovery Elementary School opened in 2015, becoming
one of four K–12 zero energy schools at that time. It was the largest zero
energy school in the United States. The school is 98,588 square feet and was
built at a cost of $30.8 million for the building and site work, or $312 per
square foot.[2]

This per square foot cost excludes the cost of the solar system to enable
consistent comparisons among the case study schools considered in this book,
noting that three of the case study schools used third-party investors. The
Discovery solar PV system cost $1,510,000 for 496 kW or $3,044 per kW. In
comparison, the solar PV installed at the Northland Pines schools in 2017 cost
$1,519 per kW.[3] The considerably lower installed cost per kW at Northland
Pines compared to Discovery presumably reflects further declines in installed
solar costs between 2015 and 2017 and different solar market conditions in
the greater D.C. area versus in Wisconsin.

The Discovery Elementary cost, at $312 per square foot, initially appears
high when compared to the $116 per square foot cost at Northland Pines High
School, but that cost difference reflects two important factors. First is the
considerable construction cost inflation between 2006 and 2015. The second
factor is the location in the D.C. area with higher construction costs relative
to the northern Wisconsin location.

The regional school construction cost average for the area including Virginia, West Virginia, Delaware, Maryland, and Washington, D.C., was $237
per square foot in 2014 for elementary schools, placing Discovery Elementary
above the regional cost average.[4] A construction site in a major urban location

such as the D.C. area will usually result in higher construction costs than smaller cities and towns that are well removed from large urban metropolitan areas, thus understandably pushing Discovery's costs above the regional cost average.

Adding to the D.C. area cost pressure for Discovery Elementary were site constraints. A U.S. Department of Energy case study of Discovery Elementary provides a useful description of the site.[5] It notes that the site was adjacent to an existing middle school and thought to be unusable for an additional school due to a sloped hillside. An aerial image provided in the DOE case study shows how the school was fit in a tight space on that hillside. This would have increased the site costs that were $4.1 million and included in the $312 per square foot school cost.

Considering these factors, it appears that Discovery Elementary School was delivered at a competitive cost point, probably no more than what a conventional school design would have cost on that same site. The DOE case study drives this point home by noting that a comparable school built in nearby Prince George's County, Maryland, in 2014 cost $294 per square foot. The cost information reinforces an important theme and goal in developing zero energy sustainable schools—namely, that sustainable solar schools can be delivered in the marketplace at no more cost, and sometimes lesser costs, than what is typically being spent on conventional schools in their region with similar sites. Discovery Elementary joins Northland Pines High School and River Crest Elementary School as well as other case studies in this book in providing evidence of sustainable schools being competitive or more than competitive on school construction costs.

The original sustainable goals at Discovery Elementary included energy efficiency and LEED Silver certification. Architect Wyck Knox of VMDO Architects suggested that the school could be a zero energy school for the same cost. The school district decided to test this concept by proceeding and bidding the construction for a zero energy–ready building and including the solar as an alternate bid. An alternate bid is for a project element for which bidding is sought, but that may or may not be included in the final project. In the end, the alternate was included and the cost of the zero energy building and the solar came in $1 million under budget. LEED certification under Version 3 was at the gold level.

The Arlington and Northland Pines School Districts share interesting similarities including the following:

- They were focused on creating optimal learning environments, using the schools as living laboratories.
- They set high goals for sustainability from the beginning of the project and exceeded these goals.
- They had to contend with a constrained budget, at or below comparable conventional designs for that location.

- They engaged design and construction teams with experience in sustainable schools, and in the case of Discover Elementary, in solar, zero energy school design.
- The school design processes relied heavily on detailed energy modeling.
- They exceeded their initial goals, including the initial goal of LEED Silver certification.

The two school projects also have some important dissimilarities. An important dissimilarity is that the serving electric utility in Virginia has a high net metering threshold compared to the much lower threshold at Northland Pines. The net metering threshold in Virginia is 500 kW, which Discovery Elementary used to the maximum by installing 496 kW of solar PV.

Other dissimilarities include the following:

- A more moderate climate at Discovery Elementary
- An urban site for Discovery Elementary versus a rural site for Northland Pines
- A large, flat site at Northland Pines versus a small, sloping site at Discovery Elementary
- An initial explicit zero energy goal at Discovery Elementary

The notion of zero energy was not something even considered in 2003 when the Northland Pines project was being started, but the school district, frustrated in its inability within its budget to include sizable renewable energy at the start, anticipated adding renewable energy in the future.

Forming and working with effective project teams is an important factor in the success of these schools. At Discovery Elementary, the Arlington Public Schools Assistant Superintendent for Facilities and Operations John Chadwick was instrumental in selecting the architectural firm VDMO, which thought the school would be a candidate for zero energy.[6] VDMO partnered with CMTA Engineers, which had previously done the engineering design for Richardsville Elementary School in Bowling Green, Kentucky. These building professionals worked with the school district in determining what was achievable and empowered the school district in expanding the sustainable goals of the project.

Discovery Elementary School's architect Wyck Knox encountered an interesting example of cost shifting during the design process. *Cost shifting*, or value trading, is a process in design and construction where costs savings in one area of a project are used to fund added costs in another project area. When solar costs were much higher a decade ago, it was common to see cost shifting on school projects to save enough to add at least a small solar PV system.

As the Discovery Elementary project team sought opportunities for energy efficiency, they considered going with triple-pane windows rather than double-pane windows to increase energy efficiency. The cost of adding triple-pane windows came in at $119,000. The cost of upsizing the solar PV system to provide the incremental energy required by staying with double-pane windows instead of shifting to triple-pane windows was a mere $9,000.[7] That was an easy trade-off for the project team and freed up funds for other productive use.

Some of the sustainable features include using CO_2 monitors to adjust the volume of outside air according to occupancy in various rooms. This measure reduces energy use while maintaining high indoor air quality. Daylighting is provided with solar tubes as well as windows to reduce electric lighting requirements and to provide a more attractive indoor environment with views. Careful attention was paid to orientation and building shell. The school used a geothermal HVAC design using ground-sourced heat pumps, which avoids the use of natural gas entirely. The result is a remarkable Site EUI of 15.4 $kBtu/ft^2$ per year that CMTA claims is the most energy efficient school in the county.[8] The Source EUI is 48.8 $kBtu/ft^2$ per year.

Energy use for the year June 2016 to May 2017 was 442 MWh. The on-site solar system provided 543 MWh for an annual net export of 101 MWh. Noting the drop-off in performance of heat pump systems in colder weather, along with lower solar production, the only months with net use of utility power were December through February. This is a powerful demonstration of the feasibility of on-site solar PV in schools to provide all, and even significantly exceed, a school's energy requirement with solar on a net-zero or, in Discovery's case, a net-positive basis.

As the district and its project team managed the project to include the unplanned solar and stay within budget, the Arlington Public Schools District did not turn to third-party investors to fund the solar and leverage tax credits in their effort. The solar PV system cost $1.5 million for the 496-kW solar PV system, out of a budget of $33 million. An interesting question is, What might have been further gained financially by leveraging the federal tax credits depending on what financial terms may have been offered? If a third-party approach had been used, how would the $1.5 million saved been used or would the school district have preferred to reduce the project budget?

An important lesson from the Discovery Elementary school project is the value of a deliberate management process by the school district and its project team. The school district started out with a clear, sustainable goal inspired by a nearby LEED Gold elementary school. In early discussion with their architect, however, they amended that goal in the early going to include solar within the same budget. They selected an engineering firm with experience with zero energy schools and worked as a team through early project design guided by energy modeling. All through this process they kept the budget in focus and ultimately met that budget with more than enough solar PV capacity for a zero energy school.

This is the equivalent of a river trip where almost everything exceeds expectations. The plans are well laid, the participants all pull together with their schedules and gear, the river flows are as expected, and the weather is great. The team on the trip makes good reads on the rapids, hitting the entries and avoiding the holes.

Discovery Elementary does pose an interesting additional financial optimization question beyond the question of whether it would have financially benefited by one of the third-party investment approaches. Discovery Elementary is clearly saving on energy costs and the DOE case study notes this. The interesting question, however, is, Could the energy cost reduction be improved by the addition now or in the future of battery storage?

The optimization challenge at any school is finding the optimal PV system size and battery capacity, along with the most financially beneficial path for procuring these systems. In the case of Northland Pines, the solar systems installed in 2017 were deliberately limited to the optimal point financially using TPI so that the PV solar would be cash flow positive from the beginning. As battery cost declines, a further investment in batteries and solar is anticipated, but perhaps not all the way to zero energy if a cash flow positive criterion is used.

Without running a financial analysis for Discovery Elementary, it can't be determined whether battery storage would improve the financial performance to justify procurement or lease. Battery storage and perhaps other building operating adjustments might be added to further reduce utility costs and add revenue for grid support paid by the ISO (independent system operator). The question of battery storage at Discovery Elementary is a less pressing question than at schools in locations with low net metering thresholds.

School districts in general will face these types of financial optimization questions. At a given point in time, districts face school project decisions and make the calls as to the best approach from the mix of technology and financing options available. These choices will be made, noting that future options may become available.

Discovery Elementary is an outstanding example of a sustainable solar school that was delivered on a competitive budget.

NOTES

1. "Discovery Elementary School," American Institute of Architects, 2017 Cote® Top Ten, https://www.aia.org/showcases/71481-discovery-elementary-school-.

2. "Zero Energy Is an A+ for Education: Discovery Elementary," U.S. Department of Energy, Building Technologies Office, Zero Energy Case Study, August 2017.

3. Hoffman Planning, Design & Construction, internal documents from Solar Planning Services for the Northland Pines School District solar project, 2017.

4. "20th Annual School Construction Report," *School Planning & Management*, February 2015.

5. "Zero Energy Is an A+ for Education: Discovery Elementary," U.S. Department of Energy, Building Technologies Office, Zero Energy Case Study, August 2017.

6. Cyndy Merse, Net-Zero Discovery Elementary in Arlington, VA Raises the Bar for Energy Efficiency, Green Schools National Network, Jan 13, 2016.

7. Ken Edelstein, "A Wave of Net Zero Energy Schools Crests in the South," New Buildings Institute, Kendeda Fund, April 11, 2017.

8. CMTA Engineering website.

Chapter 13

West Coast Middle School

Jeffrey Trail Middle School is a 75,178-square-foot school that was completed in 2013. It is in Irvine, California, part of the greater Los Angeles area and a bit south of Disneyland. It is part of the Irvine Unified School District. Its climatological, regulatory, and community conditions stand in sharp contrast to Northland Pines in almost every respect possible. It is also quite different from Discovery Elementary. The Los Angeles region has more than 18 million people. Eagle River has fewer than 1,500 people. The greater DC area falls in between.

The Jeffrey Trail Middle School is one of the schools in the California Proposition 39 Zero Net Energy Schools Pilot Program.[1] As of 2016, the school reported a Site EUI of 29 kBtu/ft^2 per year and provided on-site solar of 12 EUI for a net Site EUI of 17. The overall Source EUI was not reported but would be expected to be roughly 70–80 kBtu/ft^2 per year. The school is a good way to zero energy, which is referred to as *zero energy capable* and was one of the case studies featured at the fourth annual Getting to Zero National Forum held in April 2018 in Pittsburgh, Pennsylvania.[2] NBI (New Buildings Institute) and RMI (Rocky Mountain Institute) were hosts for the forum.

The contrasts in climate between Northland Pines and Jeffrey Trail are like comparing conditions on the Wolf River or Flambeau River in Wisconsin to the Green River near Moab, Utah. For starters, the Wolf and Flambeau will be entirely ice- and snow-covered in winter, whereas the Green can still be run from March into late Fall. The former rivers run clear while the water in the Green River appears like thick, brownish pea soup. The Green River is surrounded by canyon walls and an arid, mostly treeless landscape. The Wisconsin rivers are enclosed by verdant, continuous forests and modest hills in some locations.

In contrast with Northland Pines and Discovery Elementary, with high humidity for part of or much of the year respectively, Jeffrey Trail is dry.

111

The ASHRAE climate zone identifies it as warm and dry. As a result, Jeffrey Trail Middle School has modest requirements for heating, need for cooling, and essentially no concern for dehumidification.

The regulatory contrasts are also great. California has a higher net metering threshold at 1,000 kW and allows PPAs (power purchase agreements). The school district established a power purchase agreement with SunEdison. As noted earlier, the legal status of PPAs in Wisconsin is uncertain and for the time being no PPAs are being used. Virginia has a higher net metering threshold as well, although not as high as California's. Discovery Elementary used the higher net metering threshold to maximum advantage. Despite these differences, the question for each of these schools is the same. Will going to on-site solar provide a financial and educational advantage for the school?

Facing relatively high electric rates and demand charges and with the expectation that electric rates would increase at a 4.5 percent annual rate over the life of the school, the Irvine Unified School District proceeded with the PPA for solar. The 423 kW of solar PV panels are located on the parking canopy.[3] In selecting a PPA, the Irvine Unified School District chose a long-term arrangement for buying power on an ongoing basis from SunEdison, which placed the solar system at the school. The school district will not own the solar systems.

This contrasts with Discovery Elementary, where the Alexandria School District purchased the solar PV system with its own project funds when it built the school. Northland Pines used another form of TPI that operationally is somewhat of a hybrid of the Discovery and Jeffrey Trail approaches. Until Northland Pines exercises its buyout option at year 12, Northland Pines will have monthly payments that financially will work much like monthly PPA payments at Jeffrey Trail. At that time monthly payments are no longer made, and the situation becomes like Discovery with no monthly payments for the solar. At that point, the cash flow jumps at Northland Pines from slightly positive to strongly positive. Both the Discovery and Northland Pines school districts will then be responsible for operations and maintenance. Jeffrey Trail has no operation and maintenance responsibility.

Jeffrey Trail Middle School, along with the rest of the district, is planning to add battery storage as part of an incentive program by SCE (Southern California Edison) to reduce peak loads on the grid. The storage program allows both the storage of solar power and the storage of power purchased from the grid during off-peak periods. The power stored in batteries will be used to avoid or reduce grid purchases during peak-load times when the cost of power is the greatest and to add power when most beneficial for the entire power system. This is the smart grid / smart school model that will make its way across most of the country in the coming years.

David Bell of PJHM Architects describes a design process at Jeffrey Trail that is common for the Irvine Unified School District. "The architecture does not drive the education. The education drives the architecture."[4] This description of a design team eliciting educational and program needs in the early planning process to guide the design process is consistent with the collaborative design process at both Northland Pines and Discovery Elementary.

The design process utilized the Collaborative for High Performance Schools (CHPS) performance standards for exemplary energy and water efficiency. The school is designed as a single building for all classrooms, labs, administrative offices, and other spaces—except for the gymnasium that is a stand-alone building. Combining spaces, excluding the gym, into a single structure reduces energy losses with the outdoor environment by reducing the surface-to-volume ratio and provides for better security. Northland Pines and Discovery Elementary follow a similar approach in that both of those schools are in a single building.

All these schools share similar sustainable design features in terms of using daylighting, electric lighting designs that are low power density, use motion sensors to control lights, and in some cases, use daylight sensors that measure the amount of daylight present to control the electric lights. Other measures common to these schools reduce plug loads and IT loads. *Plug loads* refers to miscellaneous loads that occupants may add including smart screens, computers and handheld devices, coffee makers, small refrigerators, and vending machines. Jeffrey Trail Middle School's building control systems are unique among these three case study schools—not in function, but in terms of how they are controlled, which is by district facilities management staff at a central location.

The HVAC system design at Jeffrey Trail takes a different direction than Northland Pines and Discovery Elementary in terms of using somewhat less-efficient, packaged gas/electric rooftop units in place of the central boiler and chiller plants at Northland Pines and a central geothermal system at Discovery Elementary. Packaged rooftop units provide cooling through compressors that are part of the unit and provide heat through natural gas combustion. While less efficient than central systems, they have the advantage of lower first cost. With the reduced need for heating and cooling in Irvine, the loss of efficiency is less of an issue.

The strategy at Jeffrey Trail was to provide greater flexibility in controlling different parts of the school. Rooftop units serve a distinct area such as a classroom wing, the central library or media center, and a gym. Rooftop units are located to serve areas of the school that operate together and can be turned to unoccupied modes or even shut off when that area of the school is not in use. The dry climate allows for reducing ventilation for periods of time while controlling for mold, which is much more an issue in the humid climates of the Midwest and East Coast.

The combination of lower-cost mechanical systems, the use of PPA for solar, and other cost-savings measures enabled the Irvine Unified School District to provide a solar school at a construction cost quite comparable to regional cost averages. The school was contracted at $21.1 million, which comes to $289 per square foot.[5] The regional cost averages for middle schools for the years 2012–2014 averaged $261 per square foot.[6] What is remarkable about the Jeffrey Trail cost is that it is for a school in Los Angeles, which one would expect to have a significant cost premium above the regional cost averages.

The solar PV system at Jeffrey Trail is located not on the school roof or ground, but on parking canopies. Using solar panels as parking canopies provides an additional benefit in the form of shade and rain shelter for vehicles parked underneath. As the proportion of electric and hybrid vehicles increases, shading becomes a more valuable service as it keeps vehicles and the batteries they contain cooler, extending the life of the batteries—not to mention other components, including painted surfaces. It also reduces energy use for staff and visitors to cool their over-heated vehicles. The rain and snow shelter benefit is limited in that there are few rainy days, but would be more valuable in other locations in the United States. The solar capacity is 423 kW and is part of the Irvine Unified School District's total solar capacity of 5.3 MW.

Jeffrey Trail Middle School provides a contrasting example of a zero energy approach for a school compared to the two previous case study schools. The contrasts include the use of conventional, lower-cost, rooftop units rather than the ground-sourced heat pumps for geothermal or central boiler and chiller systems. If natural gas use and its associated carbon footprint are to be eventually offset, the school will have to become a net exporter of electricity to achieve full zero energy. The solar capacity included so far is considerable and results in very low remaining energy use.

The use of a PPA for the carport solar results in ongoing cost savings, no up-front cost, and no maintenance responsibilities, which are provided by SunEdison and any subsequent owner of the PPA. Plans are to add batteries, which would provide further cost savings for Jeffrey Trail and for the serving utility. It will be interesting to see if further solar is added at the time batteries are installed. The school district could take the school to zero energy at the time batteries are installed or at some future time. The on-site solar could potentially be added with more covered parking or with rooftop solar.

Jeffrey Trail Middle School is an exemplary example of solar energy and sustainability. It is also an ongoing case study pending future developments with battery storage and additional solar. It has the potential to evolve from a very low-energy school with solar to a zero energy school. The climate and regulatory environments make this an easier transition than at Northland Pines. Like Northland Pines, however, it would have to become a net exporter of electricity to offset natural gas use or replace its mechanical systems to ground- or air-sourced heat pumps to become zero energy. The timing and events in its evolution will be determined by the opportunity for further financial gain and leadership within the school district.

NOTES

1. "Ultra-Low Energy School Case Study: Jeffrey Trail Middle School. New Buildings Institute," Getting to Zero National Forum, April 17–19, 2018, Pittsburgh, PA.

2. Ibid.

3. Ibid.

4. Ibid.

5. California Department of General Services—Division of State Architect, "Application Summary," December 14, 2007.

6. 18th, 19th, and 20th Annual School Construction Reports, *School Planning & Management*, February 2013, February 2014, and February 2015.

Chapter 14

Old Elementary School in the Sierra Nevada Foothills

Schools come in all ages and sizes. If zero energy sustainable schools are going to begin to have an appreciable impact on the more than 130,000 public and private schools in the United States as we approach 2030 and beyond, there will need to be a lot of sustainable remodeling and retrofit projects. Having a clear vision in a new school design such as Discovery Elementary or Jeffrey Trail seems to make a zero energy goal quite achievable. Even taking an existing newer school built with sustainability in mind but initially without on-site solar, like Northland Pines High School, and bringing it to zero energy is doable, especially if the location is in a state with high net metering thresholds and TPI is available.

A small, 60-year-old school in a small community would at first impression appear to be a less promising prospect for a solar zero energy school. The shell is likely lacking much, if any, insulation. The windows are likely to have terrible performance properties and may even be single-pane glass in aluminum frames without thermal breaks. The mechanical systems are not likely to be original, but there's a good chance that whatever equipment is in place is dated, inefficient equipment that is likely not operating as designed. The building control system, such as it is, may have gone through various adjustments by maintenance personnel to make due. The net result is often poor indoor air quality due to inadequate outside air supply, mechanical noise, and perhaps discomfort, with temperatures too high or low.

Perhaps the best river and lake travel analogies are too old, inferior rafts or boats that still could be used on a trip, but at some added effort, risk, and lack of enjoyment. Boundary Waters and Quetico Provincial Park modern-day voyagers have mostly turned to lightweight, fast Kevlar canoes. As the speed of a hull increases with boat length, these canoes are sleek and long, say, 18 feet. The weight advantage at around 45 pounds is especially important for

117

portaging between lakes. *Portaging* is carrying gear and canoes from one lake to the next. The advantage of carrying a 45-pound versus, say, an 80-pound canoe over narrow, rough, hilly trails for up to a mile or two is evident.

Despite these obvious advantages, in the Boundary Waters and Quetico one occasionally runs into old Grumman aluminum canoes—slower in the water, terribly loud when the hull hits a rock, and, at 75 or 80 pounds, heavier for portaging. The usual reason for the older, inferior canoes is economic. They've long since been paid for and the canoe party, or perhaps scouting troop, does not have the cash sitting around to just buy or rent the lighter, efficient canoes. The old equipment has sufficed before and can suffice again.

The difference with old equipment in a school is that the indoor conditions are likely to be less comfortable and conducive to learning. Issues may include uncomfortable temperatures, humidity, high amounts of glare, poor lighting in terms of balance and light quality, noise from the operation of mechanical equipment, and poor indoor air quality—including high levels of CO_2, mold, and particulates.[1] Uninspiring indoor environments, including a lack of views to the outdoors, put a damper on learning and health.

Newcastle Elementary was a school that mostly fit this description of what problems might be encountered in a 60-year-old school. It had a long list of deferred maintenance items, had glare issues severe enough that some of the clerestory windows had been painted over to limit glare, and had noise issues from the mechanical wall units in their portable classrooms so that the units were sometimes shut off, so students could hear their teachers.[2] Newcastle Elementary would not seem a likely candidate for achieving zero energy.

To add to the challenge, Newcastle Elementary had a very limited budget to work with, which both limited some of the options but also serves as a powerful example for schools on very limited budgets. Despite these challenges and limitations, it became one of the case studies featured at the fourth annual Getting to Zero National Forum held in April 2018 in Pittsburgh, Pennsylvania, where Jeffrey Trail Middle School was also featured.

Newcastle Elementary is a 31,536-square-foot school originally built in the 1950s. It is one of four schools in the Newcastle Elementary School District. Newcastle Elementary School is located in California's Sierra Foothills, about 30 miles northeast of Sacramento.

Early inspiration to improve Newcastle Elementary came from California's Proposition 39 program to develop an Energy Expenditure Plan.[3] The Newcastle Elementary School District Superintendent Denny Rush, the school board, staff, and community members worked with their project team to understand the educational, health, and energy-cost savings implications of school renovation. Examples of other ultra-low energy school projects helped in explaining the benefits.

The school district and its project team identified additional financial resources in a ZNE (zero net energy) Pilot Program managed through their utility Pacific Gas and Electric. The project team also identified third-party investors in the form of an energy services company, ABM Building Solutions, that could assist in helping the school district with the first cost challenge of its restricted budget.

The goals for the remodeling project that was completed in 2017 emerged from the collaborative effort of the school district, design team, and energy services company. They were to provide the best learning environment possible, address deferred maintenance items, reduce operating costs, and leverage outside funding. Some specific objectives included glare-free daylighting, outdoor views, adequate outside air ventilation, reduced background noise from the HVAC system, and thermal comfort.

District Superintendent Denny Rush and the project team realized that despite the age and the deferred maintenance items, the school was somewhat efficient. With a set of aggressive retrofits, controlling plug loads, and the addition of an on-site solar PV system, zero net energy was a plausible target. As in the other case studies, energy modeling provided critical guidance in designing and implementing the retrofitting of Newcastle Elementary. Modeling was provided by the energy services company ABM Building Solutions, as well as by a design consultant, Point Energy Innovations, provided through the ZNE Pilot Program.[4]

The California Prop 39 ZNE Pilot Program required a Site EUI of 20 kBtu/ft^2 per year or lower. With the guidance of the energy modeling, the project team guided the design to an anticipated Site EUI of 14 kBtu/ft^2 per year, which is before considering the contribution of solar energy. The Source EUI is not reported but would be anticipated at about 40 kBtu/ft^2.

The retrofit measures included replacing old fluorescent tubes with LED lights. New light fixtures and modern controls rather than just new lamps would have been nice, but the budget couldn't afford that cost. The existing daylighting systems were recovered and adjusted using lower visual transmittance glass and raising the exterior awnings. Daylight into interior spaces was added with light tubes.

Wall mounted mechanical units in the five portable classrooms were replaced with much higher efficiency units with much reduced noise levels. A high-efficiency rooftop unit was installed over the multipurpose room. A BMS (building management system) was installed to control and monitor the HVAC units.

Based on energy modeling and allowing for a 20 percent margin, a 108 kW PV system was installed that was estimated to provide 150,400 kWh per year. Roofs on existing buildings are often sufficient to support the relatively light loads of solar PV systems. Even ballasted solar systems that use blocks to hold down the racking and panels typically weigh less than 5 pounds per square foot. The existing roof at Newcastle, however, would not support a solar PV system, thus a ground-mounted system was installed.

Working through all the unknowns of retrofits and remodels is a challenging task for districts and their project teams. Unknowns include hazardous materials abatements for contaminants such as lead paint and asbestos. The Newcastle school district and their project team deserve tremendous credit for improving the school's learning environment in the context of a zero energy retrofit.

This outcome took place in a circumstance where there were fewer economies of scale in a very small project in purchasing equipment and construction materials, and for staging construction crews and equipment. For example, procuring a 500 kW solar PV system is going to achieve lower pricing than a smaller 108 kW PV system. Similar economies-of-scale cost advantages were missing on everything from the HVAC equipment and controls to the solar tubes and lighting retrofits.

What is powerful about the Newcastle Elementary example is that it serves as an example of a very small school district facing significant operational and deferred maintenance needs with a very limited budget. It was also facing other priorities as it worked through its remodeling and retrofit decisions with its project team. This is a situation that is being played out in hundreds of examples across the nation every year in small and larger schools.

The school district working with its project team developed a sustainable project that it could afford to implement. It might be better stated that the Newcastle Elementary School District could not afford to miss out on the opportunity in front of it. Most importantly, it provided a very significant improvement to the educational and classroom environment with better daylight and lighting, improved views, better thermal comfort, and much lower noise levels coming from the HVAC equipment. A zero energy approach supported by solar was also the best financial approach.

Newcastle Elementary is listed on the NBI emerging zero net energy list. This means that the intention at Newcastle is to achieve zero net energy, and time will tell whether it provides the 12 months of operational data to verify achievement of this goal. If the estimate of having an extra 20 percent margin in the design of its PV system is accurate, Newcastle should be on the NBI Zero energy list shortly. Seemingly small decisions such as shutting off unused equipment and unused lights will be some of the attention to detail required to fully achieve the goal of zero energy and the financial savings that come with it.

As we watch Newcastle's performance over the coming years, it serves as an outstanding example of what should and can be done on a limited budget in a small, older school.

NOTES

1. The EPA has considerable materials on the web related to environmental quality and schools—for example, "Improve Academic Performance through Better Indoor Air Quality in Schools" U.S. Environmental Protection Agency, https://www.epa .gov/iaq-schools/improve-academic-performance-through-better-indoor-air-quality-schools.

2. "Emerging Zero Net Energy School Retrofit Case Study: Newcastle Elementary School," Getting to Zero National Forum, April 17–19, 2018, Pittsburgh, PA.

3. Ibid.

4. Ibid.

Chapter 15

Southern Elementary School

Richardsville Elementary School located near Bowling Green, Kentucky, is recognized as the first, full-sized K–12 zero energy school in the United States. The 72,285-square-foot school was substantially completed and occupied in September 2010 and is in the Warren County Public School District. It is on the New Buildings Institute's ZNE (zero net energy) verified list.[1]

Like the other solar school case studies selected for this book, which were at or below conventional school costs for comparable sites, Richardsville Elementary was built at the regional cost average for the region, encompassing Kentucky, North Carolina, South Carolina, and Tennessee. The project cost was $14,927,000 or $207 per square foot. This cost included the cost of the solar PV systems that were installed as part of the new construction project.

Had the solar systems been purchased five years later, its solar PV system cost would have been dramatically lower. The solar system consisted of 208 kW of thin film PV located on the roof and 140 kW of crystalline PV panels placed on a parking lot shading structure. The PV system had a budget of $2.8 million, resulting in a cost per installed kW of $8,046.[2] Noting the $1,484/kW cost at Northland Pines in 2017, the dramatic decline in installed solar PV systems costs since Richardsville was completed in 2010 is evident.[3]

The cost of Richardsville Elementary school without its solar system was $12.1 million or $168 per square foot. The regional cost data for elementary schools in the region was $165 per square foot in 2009, $145 in 2010, and $185 in 2011. Kenneth Seibert from CMTA, Inc., the engineering firm that provided the mechanical and electrical design and the energy modeling for this exceptional, early, zero energy school effort, writes that the budget, without the solar system, fit within the typical budget assigned by the Kentucky Department of Education.[4]

In river-running terms, managing to include solar PV with the extraordinary high cost of solar in 2010 in Richardsville Elementary is like facing an extraordinary challenge on a river. An example might be facing flood-stage conditions on a river, which changes the character of the river. Some rapids that may have been routinely maneuvered under normal water conditions become extremely challenging at flood stage. Flooding often moves sand, gravel, boulders, and trees that have been washed in the river. The river no longer reads as might be expected, and there may be unanticipated obstacles, including sweepers—trees or tree trunks that have been caught crosswise to the river, affording no way around them. That's a problem when they suddenly appear around a bend in the river with insufficient time to get to shore to portage around the obstacle.

The school district and its project team knew what they were facing, but decided to run the river under those conditions, rather than wait a few years for solar prices to possibly decline. Things often have clarity in hindsight, but betting on a dramatic, near-term fall in solar prices perhaps did not appear to be a real good near-term bet. In any event, there were surely other important educational and community issues driving the timing of Richardsville Elementary.

The school district and the project team were coming off the earlier success at the Plano Elementary School also in the Warren County Public School District. Plano Elementary, completed in 2007, was the most efficient school in Kentucky.[5] It did not, however, include solar energy. Following that success, the school district and its project team seem to have been quite excited by the challenge.

If a river group has cleared their schedules, has a permit for a certain date, and is deciding to go ahead on a new trip, it will judge the situation and decide. And the challenge of high water and different river characteristics can incite the adventurous spirits.

In Kenneth Seibert's telling of the story, Mark Ryles, facilities director for the Kentucky Department of Education, posed the question, "How would one design a net zero energy school and how much would it cost."[6] Coming off the success of the Plano Elementary project, the Warren County Public School District decided to proceed with its ambitious new project at Richardsville. The new project would again incorporate high performance levels of energy efficiency and now add sufficient solar energy to reach zero energy.

Like the other case study schools, the school district had formed an effective project team at the beginning of the project. The team included Sherman Carter Barnhart as architect and CMTA for engineering, the same firms that had worked on Plano Elementary. The school district, including the school facilities manager, district administrator, and board members, and its project team also worked collaboratively with state regulators and utility companies as the design approach required some waivers from state regulators.

The team used an integrated design approach with heavy reliance on energy modeling in the design process and a rigorous commissioning process. Because the solar system was so costly at the time of the project, the project team had to push the efficiency of the school design particularly hard and delayed the bidding on the solar systems until as late as possible. The solar system became fully operational in January 2012.

The solar PV installed cost was budgeted at $2.8 million for the eventual 348 kW system, or close to $8 per watt. It was evident that the solar system at Richardsville Elementary was not going to pay itself off in a 15-year period as required by the Kentucky Department of Public Education.

The project team proposed a novel approach. It would use a combination of energy cost savings from solar generation and from other efficiency measures in the overall project to achieve the net operational savings to pay for the solar that would bring the project to zero energy. The school district was assisted in procuring the then high-cost solar PV system by the Tennessee Valley Authority, which offered to buy the solar generation at a substantially higher price in exchange for the solar renewable energy credits (RECs) that went to TVA. This story is described in detail in an article by Kenneth Seibert in *High Performing Buildings*.[7]

Important design elements included a high performance building shell using insulated concrete forms (ICF), a geothermal heat pump system, and a dedicated outdoor air supply system. HVAC, lighting, and even kitchen equipment were kept to minimum capacities. The school district was able to work with kitchen staff on alternative methods for preparing meals on site that reduced energy use in the kitchen. A low-lighting power density based on high performance fluorescent fixtures was implemented. The LED lighting now available at low cost would have made that task easier.

Other innovative design measures included the CO_2 control of outside air supply to building spaces. The CO_2 control matches building loading in terms of people with the necessary supply for air quality. The Richardsville Elementary is quite remarkable for the careful mix of measures that were brought to the final project design. Given the aggressive efficiency target of a Site EUI of 18 kBtu/ft^2 per year, the project team had to think out of the box. The NBI reports later data at a Site EUI of 19 and a Source EUI of 59.9.[8]

The solar PV system consists of two systems. The 140 kW of crystalline PV panels are used as part of a parking shade structure. The 208 kW thin-film PV system is located on the roof. The solar systems were sized to match the school's energy requirements. Data from the 2019 NBI list show that it was a slight exporter of energy at –2.6 kBtu/ft^2 per year. Data from 2014 show that it was a slight importer at 1.30 kBtu/ft^2 per year. Slight variances in things such as plug loads, lighting use, and weather can tip the balance in any given year for a building where solar energy supply and demand are so closely balanced.

Richardsville Elementary is a remarkable success as the first zero energy school of significant size on a conventional budget. With the advances that have occurred since 2010 in solar PV technologies and cost, LED lighting for internal and external applications, and improvements in controls, the challenge of zero energy in a school from an operational and a first cost basis has gotten much easier.

Without the collaboration of TVA to provide a healthy revenue stream for the solar generation, however, Richardsville Elementary would have faced the same dilemma as Northland Pines in that Kentucky, like Wisconsin, has a low net metering threshold. The financially feasible option if Richardsville had been built without the collaboration of TVA would have been to limit the solar capacity to what penciled out financially. Additional solar would have waited until solar PV system prices had dropped. Third-party investment options may have emerged with the decline in solar costs, thus allowing Richardsville Elementary to avoid the upfront cost of adding solar.

As Richardsville Elementary stands, it is a remarkable pioneering effort in zero energy schools. It had a vision, assembled a highly capable team, collaborated with outside players—including the TVA and the Kentucky Department of Education—and moved boldly and successfully ahead to zero energy in a sustainable school. It's a project that chose zero energy without any fossil fuel use in that it adopted the all-electric geothermal design. Despite the extremely high cost of solar PV systems at the time, it purchased the solar within its limited budget.

It's remarkable that Richardsville Elementary, despite the challenge of a strict budget limit and high cost solar PV system at the time of construction reached zero energy. It joins the four other case studies in demonstrating a vision and the will to implement a zero energy solar school or a solar school that is capable of and on the path to zero energy. A powerful theme that is common to the five case studies is that these solar schools were planned, designed, and constructed at or below conventional construction cost.

The obvious point is that as these schools were provided to their districts, about half of the schools of all kinds being built at the time were above conventional design cost averages. It makes one ask what inspired these innovative school districts, their administrators, and their project teams that led them to pursue zero energy solar as a goal while simultaneously motivated to deliberately manage project cost to levels no more than conventional.

It's not so much a matter that these projects could be done on a financially competitive basis to conventional schools, but that these school districts and their project teams recognized that they had the wherewithal in their decision-making to deliver financial performance while attaining their vision of a high performing, sustainable solar school. It's these examples that are eroding the myth that green, sustainable, and even solar zero energy schools must cost more.

The final two chapters consider the implications of this recognition and the ongoing developments in technology and climate change for future schools.

NOTES

1. *Getting to Zero Status Update and List of Zero Energy Projects*, New Buildings Institute, January 23, 2018.

2. Kenneth L. Seibert, "Achieving Net Zero: Case Study—Richardsville Elementary School," *High Performing Buildings*, Fall 2012.

3. The author is Director of Sustainable Services at Hoffman, which designed and built the Northland Pines High School and later led the solar planning and financing effort for the solar installation in 2017.

4. Kenneth L. Seibert, "Achieving Net Zero: Case Study—Richardsville Elementary School," *High Performing Buildings*, Fall 2012.

5. Kenneth L. Seibert, "Small Step, Big Savings: Case Study—Plano Elementary School," *High Performing Buildings*, Fall 2009.

6. Kenneth L. Seibert, "Achieving Net Zero: Case Study—Richardsville Elementary School," *High Performing Buildings*, Fall 2012.

7. Ibid.

8. *Getting to Zero Status Update and List of Zero Energy Projects*, New Buildings Institute, January 23, 2018.

Section 4

WHAT LIES AHEAD

Chapter 16

Technological and Economic Forces

Building a zero energy solar school in the period leading up to the 2030 Challenge will benefit from, but will also be challenged by, rapid technological innovation and evolving market forces. Some of this can be anticipated and some of it will be unanticipated and unpredictable. Expect the unexpected. School district and private school administrators, teachers, facility staff, boards, and their project teams that are building new schools will have to make decisions in this rapidly changing environment. They will benefit from the experience of existing solar schools and the growing number of solar schools that will emerge between now and then.

River trips have some unpredictability—and hence the need for trip teams to be prepared to make decisions and read rivers without the benefit of fore-knowledge. In our extensively explored, remotely sensed, and mapped world, this unpredictability can still occur in many ways, including:

- Rivers that have yet to be run
- Rivers that have been run perhaps thousands of times but that are having unprecedented conditions, such a flood stages or drought conditions
- River trip groups that could have gathered information ahead of the trip but failed to do so and are running a river that is known, but not known to the group

A book that is an example of the first situation recounts Teddy Roosevelt's first descent of a river in the Amazon.[1] It takes place after Teddy has completed his second term and failed in his comeback under the Bull Moose Party. If you want fear, terror, and desperation, this is a fine read.

A preposterous example of the second situation is the running of the Colorado River through the Grand Canyon in a dory under extreme flood conditions. The run is by three river guides highly experienced with the river. This took place in 1984 as the newly constructed Glen Canyon Dam at Page, Arizona, was filling to create Lake Powell. Kevin Fedarko's *The Emerald Mile* provides a page-turner account of this remarkable descent.[2]

The third situation happens all the time, particularly with less seasoned groups. While these groups don't need to be in the situation of reading the unknown, they are. And if they are experienced enough, they can manage this situation nicely, and indeed it can add to the pleasure of needing to rely on your knowledge and experience of exploring a river being first encountered by a group of friends.

As a growing number of solar schools are built, school districts and their project teams will continue to encounter new situations. School designs are unique in most situations. This is in part due to the unique needs of each community planning a school, ever-changing technologies and materials used in schools, and tradition in the design and construction community of not building cookie-cutter schools. Thus, school buildings have not gone through the mass marketing and homogenization process that characterizes such products as motor vehicles, appliances, and to some degree home designs. Schools continue to be customized to the specific needs of a community and school district.

The exceptions are some modular schools or classrooms that are gaining some traction in the school market. Another exception occurring in some traditional design and construction is the repetition of designs in some rapidly growing school districts that do repeated, nearly identical projects of, for example, elementary schools. The ongoing, rapid change in technologies, costs, supply chains, and financial conditions serves to counteract the interest in replicating traditional and modular school designs.

Some modular classroom and school manufacturers have responded to sustainable technologies. A solar modular school or classroom is feasible, whether the solar is provided on the roof or on adjacent parking structures or land. A zero energy solar classroom called Gen7 Zero Net Energy—Solar Classrooms is now available on the market.[3] This design and other modular designs are being manufactured in California by American Modular Systems.

A dramatic example of rapid change in school design is the adoption of on-site solar PV systems. Until about 2010, the use of large on-site solar systems that would provide for more than a few percent of a school's energy requirements was almost unheard of. The cost was simply too high, even in areas where utility rates were high. On-site solar applications at schools were typically smaller demonstrations funded in part or whole by donors, including some investor-owned utilities. Some small schools or school additions could do early zero energy demonstration projects because of small scale—say, a building of fewer than 10,000 square feet.

By 2015 the situation had changed as solar costs had dropped by 75 percent, net metering rules had been established in most states, and utility rates had continued to rise. In many states, school designs could suddenly include solar sufficient to provide a significant part of the energy or even reach zero energy and provide financial savings . The rapid rate of change has not subsided as solar prices continue to fall driven substantially by the massive investment in state-of-the-art manufacturing facilities in China. Greg Nemet's just released book is an in-depth exploration of the forces and players reducing the cost of solar PV systems in the global market.[4]

The solar import tariffs imposed by the United States at the beginning of 2018 only apply to PV panels and will expire in four years. The market response appears to be no more than a speed bump in the downward pressure in market pricing and the upward surge of solar PV installations and overall installed capacity in the United States.

Paralleling the fall in solar cost has been the development of third-party investment options that enable school districts and other nonprofits to bring solar to their schools without the up-front burden of bond borrowing or otherwise using district funds. Third-party investment leverages the current federal 30 percent investment tax credit and accelerated depreciation schedule for as long as they remain in place.

While some zero energy schools such as Discovery Elementary and Richardsville Elementary have been able to make the up-front investment in solar, the added cost of solar and, in some locations, batteries are beyond the reach of some struggling school budgets. With the fall of solar costs and the use of TPI enabling cash-flow-positive outcomes, increasing numbers of school districts and private schools will utilize the TPI market to find the best options to provide on-site solar.

Third-party investors will continue to contend with changing tax rules and market conditions in addition to changes in the cost of solar technologies and components. These conditions include the availability and level of investment tax credits, bank interest rates (if used by investors), and other changes in the tax code. State- and utility-level regulations as to what forms of TPI are permissible and by what entities will also evolve over time, as will net metering regulations. The trend over the last decade has been for the most part to increase TPI options for schools and other commercial buildings.

Stepping back to look at the larger picture of the electric utility grid and market, one increasingly likely future is one where electric utilities continue to transition away from fossil fuel. Some utilities have committed to transitioning to zero carbon by 2050. If that were to happen, an all-electric sustainable school could purchase all its power from the grid and be zero carbon. This development, if it blossoms, would not alter two important considerations. First, the financial and environmental advantages to zero energy solar schools built between now and 2050 would remain. Second, it may well be the case that decentralized power generation at the school site in 2050 will be less costly and more resilient in part due to less transmission and distribution cost with decentralized and dispersed generation relative to a highly centralized power grid.

In the near term, solar investment tax credits are scheduled to revert from 30 percent to 10 percent in phases for solar projects starting construction between 2019 and 2022. Following their introduction in 2005, solar investment tax credits have been extended three times with the last extension occurring in 2015. With the growing concern or even panic in response to climate change, more extensions of the investment tax credit for solar and batteries would not be surprising.

Another potential policy that would accelerate the adoption of on-site solar in schools and elsewhere is a carbon tax or a cap-and-trade system. A carbon tax would serve to increase the costs of utility power to the extent that fossil fuels are part of the generation mix.

School districts and private schools will need to follow the TPI market to see how the solar offerings continue to evolve. As climate impacts play out, the interest in mitigation measures would be anticipated to increase, resulting in future extensions and other supportive changes in incentives and tax treatment.

Even if investment tax credits for solar were reduced to 10 percent in 2022, as scheduled, and bank interest rates settled in the range of, say, 5 percent to 7 percent, the decline in solar PV system costs and stable to increasing utility rates will usually be sufficient for on-site solar to be done in various forms of TPI on a cash-flow-positive basis in most areas of the United States. This is likely to include solar PV investments where ownership stays with third parties and other forms of TPI where ownership is eventually purchased by or transferred to school districts. As new schools or major remodeling projects are considered, solar will continue to be evaluated with whatever are the options available to the school district or private school at that point in time.

On-site energy storage is a quickly developing complement to on-site solar as discussed in chapter 9. The advantage of batteries is a combination of being able to store excess solar production and use it to either offset future school power needs or selling it on the grid for other users and for its value as grid support. On-site energy storage is of particular interest in states with low net metering thresholds. It will also be of interest in states with higher net metering limits as there are likely to be emerging financial benefits for energy storage for schools beyond those associated with achieving zero energy. The Jeffrey Trail Middle School is planning for batteries and is in a high net metering threshold state.

The advance of on-site energy storage will be influenced by developments in battery technology and pricing in parallel with the development of smart grids, microgrids, and smart buildings. It's all about the ability to obtain price signals at a school in real time and being able to adjust school operation and energy storage to take maximum financial advantage.

Innovation in lithium-ion batteries is considerable. Whether this is the best battery type to achieve financial performance in a stationary setting such as a school where battery weight is not a real concern is not known. The focus of intense interest is in how far battery costs will fall, what the durability and robustness of the batteries will be, and the expansion of the recycling market for batteries. Argonne National Laboratory is leading a current effort focused on developing technologies to enhance lithium-ion battery recycling in the United States.

The transition away from natural gas is an important question, especially for schools located in parts of the country with larger heating demands. In the Northland Pines case study, natural gas costs accounted for 30 percent of the annual energy cost. The future question for schools in areas with pipeline natural gas available and relative high heating demands is whether to forgo any use of natural gas to avoid the associated carbon emissions footprint. Noting the environmental impacts associated with natural gas production, transport, and use, a strong case exists to exclude natural gas use in schools.

As demonstrated in the Discovery Elementary School and Richardsville Elementary School case studies, geothermal systems are the foremost mechanical system option to meet heating needs without natural gas. These heat pump systems are particularly efficient cooling systems as well. A second option is air-sourced heat pumps, which are less costly to install as they don't need an onsite geothermal well field, but their efficiency drops as the outside air temperature drops, especially below 10°F. Even in the more temperate Virginia and Kentucky locations, the case study school teams chose ground-sourced heat pumps rather than air-sourced ones.

Despite the cold climate and availability of natural gas in the Madison Wisconsin area, recently completed school projects and additional projects in the construction process are going geothermal. This includes a 115,000 square foot zero energy elementary school in the Oregon, Wisconsin school district that will include batteries along with a solar PV system.

The current trade-off for schools is that, because natural gas is quite inexpensive by historical standards, the school district will end up paying a construction cost premium to include geothermal rather than a mechanical system using boilers and natural gas. Operating costs will be similar to or lower for geothermal schools compared to schools with natural gas boilers and chillers (depending on the local electric and natural gas rates), but there will be a first cost premium of approximately $500,000 or more for the well field for the geothermal system for a 100,000 square foot school.

As the cost of on-site solar declines, providing less costly electricity, the operating cost advantage of geothermal will increase relative to using natural gas for heat. TPI service offerings may develop packages of on-site solar, geothermal systems, and batteries that would remove the first cost impact of geothermal systems on a school budget.

The reality is that over time school districts and other building owners will be making decisions on their new schools, school remodels, and buildings of other types in an ever-changing landscape as technology innovation occurs, regulation and tax codes evolve, and the energy marketplace evolves, including in response to carbon policies. School districts will need to collaborate with their project teams to optimize their designs in response to these conditions. Financial analysis and consideration of TPI will play an increasingly important role.

School districts and private schools, for reasons considered here and in the next chapter, will increasingly choose sustainable school designs with geothermal mechanical systems and on-site PV solar, often with batteries, as the best path forward. The TPI market has emerged as the predominant approach for providing solar and is anticipated to provide batteries in the future. TPI offerings are likely to continue to expand as market forces make sustainable solar schools ever-more desirable financially and as a platform to support education in school buildings that are sustainable living laboratories.

NOTES

1. Candice Millard, *The River of Doubt: Theodore Roosevelt's Darkest Journey*, New York: Doubleday, 2006.

2. Kevin Fedarko, *The Emerald Mile: The Epic Story of the Fastest Rides in History through the Heart of the Grand Canyon*, New York: Scribner, 2014.

3. Gen7schools.com.

4. G. F. Nemet, *How Solar Became Cheap: A Model for Low-Carbon Innovation*, New York: Routledge, June, 2019.

Chapter 17

The River Ahead

Once you've left the departure point at Lee's Ferry at the beginning of the Grand Canyon, there is no turning back until Diamond Creek at mile 226 or Pearce Ferry at mile 279. There are no landings at which the boats can be pulled off the Colorado River, except for lunch, a hike, or camping overnight. There are a few locations at which you could take a long day's climb to the rim if you cared to abandon your boats or rafts and gear. With paddles and oars only, this trip is two or more weeks, depending on water levels, and the number of side excursions for hiking, photography, and bird-watching. Barring a surprise and perhaps unpleasant event, it is inevitable that the river will be run and the destination reached.

The same can be said of a float or canoe trip on the Smith River starting at the Ranger Station west of White Sulphur Springs, Montana. Sixty-one miles are all that are required, perhaps over four or five days on the crystal river waters. The length of time depends on your desires to watch peregrine falcons, do some side hikes, or fly fish. Once you've left the ranger station, there's no turning back and no place with public access to take your boats out. The outcome, namely a gorgeous trip arriving at the Eden Bridge Campground is inevitable.

Solar schools have their own form of inevitability. Their creation is being driven by two powerful currents. One current is the growing level of CO_2 concentrations that Svante Arrhenius considered in the 1890's and have been measured through the work of Charles Keeling and others at the Mauna Loa Observatory since 1958. The source of that current is fossil energy and other greenhouse gas emissions, which are driving climate change.

Individuals, businesses, governments, and innumerable organizations—from local activists and environmental groups to international organizations—are becoming increasingly responsive to the threats and dire consequences of climate change. The consequences are evident to casual observers and documented in research data and reports. It's a visceral reaction to the emerging, widely recognized impacts on the environment and an attempt to mitigate the damage. We've gone beyond the point of *environmental protection*, if the illusion of preserving some semblance of an earlier set of environmental conditions, for example pre-industrial age, is implied by that term.

The fear of climate change is not universally held. A portion of the U.S. population continues to hold the view that climate change is either not occurring or that climate change over the last 50 years is not anthropogenic to a substantial degree. In the latter case, the view is that climate change may be happening but there is no human behavior that caused it and can now be changed to limit the trend.

For people that have recognized or are recognizing human-induced climate change, their responses have highly varying time frames. The process of adoption of contagious ideas or products moving through a population is described as a diffusion model. Some people have long since recognized and accepted that global warming and associated climate change is occurring and are acting accordingly. Others accept that climate change is occurring but aren't yet acting on their understanding of what's happening.

The response to the evidence may be like the response by farmers to hybrid seed corn in the Midwest described in Malcolm Gladwell's *The Tipping Point*.[1] There will be "Innovators," then "Early Adopters," then "Late Majority," and finally the "Laggards." Perhaps his book should have noted a small minority of people beyond the Laggards, namely the "Never Adopters."

Commercial building owners, including businesses, governments, universities, school districts and private schools, are facing choices in what they will ask for as they build new buildings and remodel others. Some will be the early adopters of highly efficient buildings and of on-site solar to achieve zero energy with those buildings. The evidence and consequences of climate change are so dire that for some of these building owners there is no choice but to go with zero energy sustainable solar buildings. No need for further studies of the cause of climate change or estimates of the economic consequences. As long as the building costs are within reason, they are committed. School districts and private schools will be operating in the same current as decision makers for other types of commercial buildings and buildings in general.

The same current is pushing other sectors of our economy. Motor vehicle manufacturers, for example, have read the current and where it's taking them. They are moving quickly to electric vehicles as reflected in Volvo's announcement that half of its cars would be battery electric models by the year 2024. Volkswagen has committed to electric vehicles. Google announced in 2017 that it had achieved 100 percent renewables in its global operations.

For people and companies not convinced of human-caused climate change and how imperative it is to act for environmental and broader or macro-economic reasons, there is a second current to consider. That current is the microeconomic realities that on-site renewable energy costs are falling below traditional fossil-based energy sources, even when the environmental costs are ignored. We've already progressed beyond the tipping point.

While Volvo's, Volkswagen's, and Ford's decisions to move to electric vehicles to help mitigate climate change are noteworthy and perhaps altruistic as corporate citizens, they are also anticipating a change in the marketplace. They can only succeed in this transition if the marketplace is going in that direction and they can profitably sell their vehicles. Their read on the marketplace is that the electric vehicle is going to out-compete the traditional, fossil-fueled internal combustion engine. Whether the logic be the greater performance, lower operational cost, or lower impact on the environment—especially when fueled by renewable energy—the choice of electric is viewed as a competitive advantage in vehicles sales and hence a financial necessity.

The strong financial current in the case for solar schools is that a sustainable solar school provides an educational platform that is more relevant and provides a better learning environment than conventional schools, can be of equal or less cost to build, and will be less costly to operate. The case study schools support this assertion. Some firms in the school design and construction marketplace have figured out how to surmount the first cost challenge, often with third-party investment arrangements for on-site solar PV systems. Other school districts that have the bonding capacity and don't need or can't access third-party investment with the financial advantage provided by the investment tax credits and accelerated depreciation are directly procuring solar PV systems. This was the case for Richardsville Elementary and Discovery Elementary. It is also the case for the new zero energy elementary school under construction in Oregon, Wisconsin noted in chapter 16 which is including batteries with the solar PV system.

The Northland Pines High School and the Newcastle Elementary case studies demonstrate that solar PV systems can be implemented after a school is built and to a school district's financial advantage. The case of Northland Pines demonstrates the feasibility of solar for existing buildings that are in more difficult areas with respect to net metering rules and meteorological conditions. Adding sufficient solar capacity to bring Northland Pines to zero energy will wait for future developments in Wisconsin utility regulations.

Newcastle Elementary demonstrates that old and smaller schools can be remodeled to provide major improvements to indoor environmental quality and the overall learning environment. Comfort can be provided while providing an energy-efficient school, and solar capacity to achieve zero energy can be provided many years after the school was originally built. There are many old schools across the United States that can learn from the example of Newcastle Elementary.

It is easier to include solar in new schools in that the school design explicitly considers solar from the beginning of the design process. The design considers roof structure, mechanical and electrical equipment location, building orientation, and specific solar designs, including parking structures, from the start. The Discovery Elementary School, Jeffrey Trail Middle School, and Richardsville Elementary School case studies are demonstrations of solar in new schools, including zero energy at two of the schools. If Jeffrey Trail Middle School adds batteries, it will be an early demonstration of the use of batteries and may provide a good opportunity to add solar to reach zero energy.

Despite the successful examples of solar schools, there will be some late adopters and a few never adopters. As the financial advantage of solar PV continues to increase, the goal of zero energy seems inevitable for many schools except for specific instances that prohibit getting fully to zero energy. Multistory schools, schools with existing roof structures that don't support solar, states that are restrictive with respect to net metering and third-party investing are among the conditions that may block fully achieving zero energy. In these cases, low EUI levels and some solar will get schools partially to zero energy.

Schools with constraints that severely limit on-site solar may look to nearby sites for locating solar PV systems and work out arrangements with their utilities for feeding solar power onto the grid. Some schools with older roofs will want to wait until reroofing is completed to avoid the added cost of removing solar to do reroofing. The bottom line is that the current of climate-change response and the current of lower solar PV cost are taking school districts in the direction of zero energy.

If the currents of climate change and financial advantage are in the direction of zero energy solar schools, school districts and private schools cannot assume that they will get that outcome by default. Just letting the school planning process move forward by pushing a new school project out into the river flow is a recipe for groundings on rocks and sandbars—or capsizing.

River runners take it for granted that a river needs to be continually assessed to verify that their boat is in a reasonable area of the river for safe travel. There may be a wide channel of flow that is suitable—or even two or more channels. In other situations, the specific location is precise with little margin for error. Missing the entry point can result in a punctured hull or raft, or a flipped boat with swim and rescue.

This raises the question of what things could get in the way of achieving a successful solar school? And what should be done to help assure success? There are a lot of things that can get in the way. The planning, design, and construction process is a multiyear process with many opportunities for missteps. The greatest barrier, however, to achieving a solar school is starting the school planning process with no clear intention or no real commitment. What are the chances of the desired outcome of a zero energy solar school if there is no clear intention? Not much.

Recognizing the value of a solar school, whether zero energy or part way there if initial conditions are not conducive for zero energy, and then setting that goal is an essential early step. As shown in the case studies, sustainable solar schools are being built at or below conventional cost levels. The early work by school district administrators and boards with their project team will need to identify realistic, but highly competitive, project costs for referendum purposes, as whatever school is built will by necessity be constructed within the referendum amount or fund-raising limit for private schools.

To help ensure its desired outcome, the school district will need to competitively solicit its project team with specific requirements specifying solar, zero energy performance, and other sustainability criteria, possibly using the Living Building Challenge, LEED, or other guidelines. It will want to do this while targeting a competitive project cost, including design and construction. While zero energy schools exist, they are still uncommon. This means that the pool of proven service providers that can be brought onto the school district's team may be limited. Even if a school district is committed to a solar school, it can't take for granted that a project team will fully understand the design implications and be able to deliver a school that meets the energy efficiency, solar, and other sustainability requirements.

While the pool of proven providers to be part of a school district's project team may be limited in some parts of the country, the pool of providers and knowledge on the topic is dramatically expanding. One need look no further for evidence of how the professional community is developing than the rapid growth of LEED professionals, the growing participation in the NBI zero energy program, the emergence of new guides like the ASHRAE *Advanced Energy Design Guide for K–12 School Buildings*, and the increasing buzz on the topic of zero energy buildings.[2] These developments are evidence that zero energy schools are arriving and the growth in professionals and resources will be there to support the inevitable future.

Many of the zero energy schools and other buildings that are being added to NBI's list are being done by professionals and firms for whom this is their first project meeting zero energy criteria. This is like the growth in the number and types of LEED-certified projects. Beginning around the year 2000, LEED certifications began appearing and the numbers grew geometrically based on the work of building owners and their project teams that were doing their first few projects.

The process of picking a project team can be fraught with uncertainty. It involves judgements by the school district administrators and boards on the experience and capability of the firms and specific professionals on the project team it is assembling. There often is a considerable range in the compensation rates and contract structures. Cost will certainly be one of the decision criteria as proposals are evaluated and compared. In the end, decisions on the project team are made and the process begins.

Group dynamics will emerge with leadership coming and emerging from the different players. As the project owner, the school district administrators, facility managers, business managers, and boards need information, education, guidance, and empowerment from their professional project team. The same statement applies to private school boards and administrators. The owners of the project in the end are responsible for the project direction and outcomes and need to be empowered in their decision-making. A clarity in the goals needs to be maintained and enforced through the ups and downs and challenges and successes as the project proceeds.

Once the broad outlines of the solar school project are identified for referendum, the referendum is hopefully passed, and contracts are put in place, the point of no return comes and passes. The school district and its project team have no choice but to make the project materialize.

A common issue for school project teams designing and building solar schools is staying current with the seemingly constant changes in many facets of the building and solar energy industries. As described in the previous chapter, the changes are technological, financial, and regulatory. Project teams will be particularly focused on developments in solar PV systems, batteries, control systems managing the purchase and selling of power to the electrical grid, other building control technologies, and mechanical systems, including geothermal and air-sourced heat pumps.

Building codes continue to advance. As a result, building code changes are pushing what could be called the worst school designs closer to being suitable for a zero energy design. Energy efficiency and solar incentives have played important roles in some states in encouraging investments. Project teams will continue to leverage these incentives where and when they are available in different forms. Evolving third-party investment options, especially for solar and battery systems, are important tools for school districts in most states.

Leadership within collaborative project teams and school districts will emerge as projects unfold, and it's likely to be distributed around the team according to the specific issues being considered at any given time. It will change depending on the specific areas of expertise relating to technology, regulatory, and financial matters. Perhaps this is like a river trip where there may be one person who is the overall trip leader. But when it comes time to repair one of the boats that's been damaged in hitting a rock in a rapid, another member of the team with boat-building and repair knowledge will lead in the repair.

In the Northland Pines High School case study, District Administrator Mike Richie provided overall project guidance through the sequence of decisions to build a new high school and to commit to LEED certification of at least the silver level, which became the gold level. Ten years later, working with members of the original project team, he led the school district through the decision process to add 418 kW of solar PV in three district schools, including 230 kW at the high and middle school. He has also laid the groundwork for a future solar project to take the high school and other district schools to zero energy. Various team members provided leadership in different areas of this ongoing journey.

A LEED professional with extensive construction management experience administered the LEED certification along with a second LEED professional with energy and environment expertise who was instrumental in performance targeting using the modeling results of yet another team member. Two project architects were key in developing the spatial form and the learning environments in response to the needs established in the pre-referendum studies and interviews. Collaboration between the architects and the construction managers kept the project on budget and on time. A commissioning agent supported the team during design and construction, after construction, and later in adding solar.

The Northland Pines School District story should provide encouragement for school districts intimidated by the prospect of doing their first zero energy school. The school district has done many firsts over time. It was one of the first LEED certifications in the state. The project team had not done a public school LEED certification before, and no one in the country had done a public high school LEED Gold certification. The professional team did have the expertise, but not all of the experience. While in 2005 and 2006 it anticipated future solar expansion, it had not at that time designed and installed large solar PV systems. The school district was sufficiently convinced to contract the project team in 2004 and 2005.

This is not to say there weren't apprehension, doubts, fears, and disagreements along the way. There were and there will be for any new school or school remodeling project, whether the school is solar or not.

In the face of these realities, school districts and private schools considering new schools and major remodels should calmly look at the currents and

- Proceed with the goal of a solar school in the context of a sustainable design that serves the educational needs of its public or private community.
- Establish clear criteria of zero energy wherever conditions permit and anticipate future solar and perhaps batteries when current conditions do not initially permit zero energy.

- Determine regional conventional school cost averages adjusted for location as needed and utilize that cost criteria and other local construction cost information in establishing a project budget.
- Build a project team that will meet these goals and criteria.
- Look for the right stuff on the team, even when this might be a first for the team.
- Collaborate as a team through the project planning, design, and construction.
- Continue to measure the results and ongoing school performance to assure the goals and criteria are met on an ongoing basis.

By design, this book should either increase confidence in school districts and private schools in moving ahead on a solar zero energy school, or increase apprehension in school districts and their team of building-industry professionals that are moving toward a conventional school design. The fear would be that the school district would be making a financial and educational mistake if it is not considering a zero energy school project at a conventional budget or even less. The climate stakes are too consequential to ignore, and the financial losses of not pursuing solar are real. With budgets for schools universally tight and challenging, why would a school district not eagerly take the lower-cost path for school construction and operation? Why not choose the best educational platform?

NOTES

1. Malcolm Gladwell, *The Tipping Point: How Little Things Can Make a Big Difference*, Boston: Little, Brown, 2000.
2. Catherine A. Cardno, "Zeroing In," *Civil Engineering*, Vol. 88, Issue 8, September 2018.

Glossary

ASHRAE	American Society of Heating, Refrigerating and Air-Conditioning Engineers.
BAS	Building automation system.
BCS	Building control system.
Btu	British thermal unit.
CAISO	California Independent System Operator.
DOE	U.S. Department of Energy.
EPA	U.S. Environmental Protection Agency.
ESA	Energy services agreement. An ESA is an agreement between investors and a school to provide multiple services that may include solar power, energy efficiency improvements, demand management, and energy education services. An ESA usually offers options to purchase the equipment provided.
EUI	Energy Use Intensity. Source EUI measures energy use at the extraction point of the energy such as a coal mine or natural gas field. Site EUI measures energy use as it enters the building. The units are $kBtu/ft^2$ (thousands of Btu per square foot per year).
Geothermal	A heat pump heating and/or cooling system where the energy exchange is with the earth using closed-loop wells.
Heat Pump	A device that transfers energy from a heat source to a heat sink. It uses a compressor cycle and refrigerant to transfer energy.
HVAC	Heating, ventilating, and air conditioning.
Inverter	An electronic device that converts direct current to alternating current.
IPCC	Intergovernmental Panel on Climate Change.
IPD	Integrated project delivery.

145

IRR	Internal rate of return. The IRR is the discount rate used in capital budgeting that makes the net present value of all cash flows from a project sum to zero. The higher the IRR, the more financially desirable is the project.
IT	Information technology.
kW	Kilowatt. It is a unit of power. It is the rate of energy use when 1,000 watts or joules of energy are used per second.
kWh	Kilowatt hour. It is a measure of energy or work equal to that done by one kW for one hour.
LEED®	Leadership in Energy and Environmental Design. It is a green certification program of the U.S. Green Building Council.
M&V	Monitoring and verification.
MEP	Mechanical, electrical, and plumbing.
MISO	Midcontinent Independent System Operator.
MW	Megawatt or 1,000,000 watts.
NBI	New Buildings Institute.
NPV	Net present value. It is a calculation of the present monetary value of a project's future cash flows.
NREL	National Renewable Energy Laboratory.
Net Metering	A billing mechanism where the utility credits electricity sold to the grid at the same price as the customer pays for electricity.
NZE	Net zero energy.
OPR	Owners project requirements.
PPA	Power purchase agreement. It's an agreement between a buyer such as a school district and a solar power firm by which the buyer purchases power at an agreed upon price from a solar system located at the school. Investors retain or sell ownership of the equipment.
PV	Photovoltaic. A solar PV panel converts sunlight to direct current electricity. PV ratings used in this book are dc (direct current).
R&D	Research and development.
RFP	Request for proposal.
STEM	Science, technology, engineering, and mathematics.
Therm	A therm is common unit for measuring natural gas and is 100,000 Btus.
TPI	Third party investment. TPI is a process where outside investors provide capital for a solar system located at a school. There are several contractual arrangements possible, including a PPA, lease, and an ESA. The TPI process benefits both the school district and the investors. The investors use investment tax credits and the accelerated depreciation thereby

reducing the cost of solar for the school. A major advantage for the school is that the solar PV system can be procured with no up-front investment.

USGBC   U.S. Green Building Council. It is a private, industry-led association dedicated to promoting green building practices.

VAV   Variable air volume. This describes a common air delivery system that provides for varying levels of air to be provided to a room in a school and can respond in real time to the heating or cooling needs.

ZE   Zero energy. It describes a school's energy use over the course of the year when the amount of energy produced on-site is equal to the energy required to operate the school. The school may import energy at times and export at other times. Zero energy is used in place of ZNE (zero net energy) and NZE (net zero energy) following the terminology of the New Buildings Institute.

ZNE   Zero net energy.

Index

Advanced Energy Design Guide
  for K–12 Schools - Achieving
  Zero Energy, 26–28, 35, 52–53, 56,
  141
aggregate installed capacity, 79
agricultural systems, 43
air-conditioning, 10
Alexandria School District, 112
American Society of Heating,
  Refrigeration, and Air-
  Conditioning Engineers (ASHRAE),
  11; *Advanced Energy Design
  Guide* by, 26–28, 35, 52–53, 56,
  141; National Technology Award,
  105
Andres, Jody, 99
Annual School Construction Reports,
  106
Apostle Islands, 70
architectural firm (VDMO), 107
Argonne National Laboratory, 135
Arlington Public Schools District, 108
Arrhenius, Svante, ix, 137
ASHRAE. *See* American Society
  of Heating, Refrigeration, and
  Air-Conditioning Engineers
atmosphere, CO_2 emissions in, 43–44
avoided costs, 72

bait and switch, 89
Barnhart, Sherman Carter, 124
BAS. *See* buildings automation systems
battery manufacturing, 39
battery storage, 15, 34, 72, 84;
  Discovery Elementary School and,
  109; of energy, 69–70, 73–75, 134;
  expected life of, 79; at Jeffrey Trail
  Middle School, 112, 134; lithium-
  ion, 135; Northland Pines School
  District and, 104; Powerpack, 74
battery technology, 134
BCS. *See* building controls system;
  building controls systems
Bell, David, 113
BMS. *See* building management system
Boundary Waters, 117–18
Bowen-Eggebraaten, Mary, 93
British thermal units (Btu), 19, 25, 28,
  32–33
budgets, 21, 100, 118, 126
building codes, 21, 142
building control systems (BCS), 11, 38,
  71
building management system (BMS),
  119
buildings automation systems (BAS), 11
Bull Moose Party, 131

149

buy-back rate, 84

CAISO. *See* California Independent
　System Operator
California: electrical rates in, 85;
　energy demands of, 69; PPAs in,
　112; school buildings in, 73; solar
　power production in, 69
California Independent System Operator
　(CAISO), 70
California Proposition 39 Zero Net
　Energy Schools Pilot Program, 111,
　118–19
cap-and-trade systems, 41, 134
carbon footprint, ix–x
carbon pricing, 40–41
carbon tax, 134
cash flow: cumulative, 82–83; of Eagle
　River Elementary School, 103;
　neutral, 58; of Northland Pines High
　School, 82; from solar systems, 81;
　from TPI, 82–83, 133–34
Chadwick, John, 107
CHPS. *See* Collaborative for High
　Performance Schools
climate change: cap-and-trade systems
　and, 41, 134; carbon footprint and,
　ix–x; CO_2 emissions and, ix, 40,
　43–44; concerns about, 134; fossil
　fuel energy and, 137–38; global
　warming and, ix, 138; human
　activity causing, 43; sustainability
　solutions for, 29; zero energy
　solar schools and, 140. *See also*
　fossil fuels; renewable energy
CM. *See* construction management
CO_2 emissions, vii, 40, 43–44
Collaborative for High Performance
　Schools (CHPS), 113
Colorado River, 131–32
commercial buildings: buyers fear
　of, 13; carbon footprint of, ix–x;
　challenges with, 13–14; demand and
　supply side of, 3; energy storage
　in, 69–70; glass selection for, 57;

mechanical systems in, 78; solar cost
　declines for, xi–xiii; split incentives
　of, 24; technologies emerging for, 4
commissioning agents, 78–79, 92–93
competitive-cost point, 40, 98, 106
construction, 14, 52, 56, 140–41
construction management (CM), 91,
　143
contamination, groundwater, 40
control systems, 78
cost-competitive sustainable buildings,
　16, 61
costs: annual average, 60; avoided, 72;
　for construction, 52; economies-
　of-scale and, 120; energy, 18, 19,
　85, 101; labor, 18; Northland Pines
　High School energy, 101, 105;
　of on-site solar PV systems, 33,
　58, 88, 135–36; operating, 23–24;
　performance and, 14–15; per square
　foot, 53; regional, 27–28, 56;
　renewable energy decline of, 139;
　Richardsville Elementary School
　solar PV system, 123; school
　buildings median, 53–54; school
　districts building, 138, 144; school
　project disparities in, 11; shifting
　of, 107; solar PV panels declining,
　33; of solar PV system, *33*, 101,
　123, 125; of sustainable schools, 15,
　56
cost-saving opportunities, 27
Cox, Tom, 99
cumulative cash flow, 82–83
cutting edge design, 9–10

Darlington Community School District,
　58
daylighting design, 63
demand side, 2–3
design-bid-build process, 50, 90–91
design process: building codes and,
　142; cost shifting in, 108; cutting
　edge, 9–10; daylighting, 63; green,
　1, 5, 98; at Jeffrey Trail Middle

School, 113; school building roof,
75; of school buildings, 28, 75, 132;
in school districts, 1, 9–10, 90–91,
136; for school projects, 51; school's
on-site solar PV systems, 132–33; for
sustainability, 53–54, 59–60, 74–75,
89–90; zero energy and, 142
diffusion model, 20, 138
direct solar purchase, *82*
Discovery Elementary School: battery
storage and, 109; competitive cost
point of, 106; project elements
of, 107; solar PV systems in, 105;
sustainability goals of, 106, 109;
triple-pane windows and, 108; as
zero energy schools, 52
District of Columbia, 83

Eagle River Elementary School, 81,
100–101, 103
economies-of-scale, 120
Edison, Thomas, 41
education, 21, 23, 62–63
electrical grid: fossil fuels ending for,
133–34; intelligent grids and, 15;
on-site solar PV systems connection
with, 68–69; renewable energy in,
38–39; solar PV system on, 1; zero
energy solar schools connection to,
28–29
electrical rates, in California, 85
electric power price, 71
electric vehicles, 138
elementary schools, 5, 25, 28, 60, 73
The Emerald Mile (Fedarko), 132
energy: battery storage of, 69–70,
73–75, 134; California demands for,
69; costs, 18, 19, 85, 101; efficiency,
24, 45; fossil-based, 12, 38–39;
operating costs of, 23–24; peak-
load time, 112; peak-use times, 71;
per capita, ix; performance, 35–36;
school buildings use of, 21, 32–34;
service agreements, 101–2; storage,
69–70, 73–75, 134; units, 18

Energy Expenditure Plan, 118
Energy Star Portfolio Manager, 18, 20,
33, 80
Energy Utilization Intensity (EUI): Site,
19; Source, 19, 25–26, 32–33; Target
Source, 28
environmental protection, 138
Environmental Protection Agency
(EPA), 18
EUI. *See* Energy Utilization Intensity

Fedarko, Kevin, 132
finance: school districts risks of, 102–3;
solar systems positive returns in,
63–64, 81; solar systems terms of,
103; sustainable schools barriers of,
77–78
firms, solar installations of, 5–6
flood-stage conditions, 124
forced inevitability, 44
fossil fuels: climate change and energy
from, 137–38; electrical grid and end
of, 133–34; energy, 12, 38–39; for
internal combustion engines, 139;
phased reduction of, 45
fracking, 40
fund-raising, for school projects, 49–50

Gen7 Zero Net Energy-Solar
Classrooms, 132
geothermal systems, 40, 57, 85, 108; in
elementary schools, 73; heat pump,
11; as mechanical systems, 135–36
Gitche Gumee, 70
Gladwell, Malcolm, 138
glass selection, 57–58, 63
Glen Canyon Dam, 132
global positioning systems (GPS), 70
global warming, ix, 138
Google, 138
GPS. *See* global positioning systems
grading criteria, of school buildings, 17
Grand Canyon, 70
Green Globes, 12, 93
green ratings systems, 12

Green River, 111
ground-sourced heat pumps, 114
groundwater contamination, 40

Hawken, Paul, 59
hazardous material abatements, 120
heat exchange, 11
heating, 19–20, 39–40, 112
heating, ventilation, and air conditioning
    (HVAC) controls, 11, 57, 62
heat pump systems, 10, 19, 40–41, 85;
    geothermal, 11; ground-sourced, 114;
    performance drop-off of, 108
High Performing Buildings (Seibert),
    125
high-performing school buildings, 25,
    31
high schools, on-site solar PV systems
    at, 4–5
Hillside Elementary School, 26
Hoffman Planning, Design &
    Construction, Inc., 59–60, 98–99
Hudson School District, 93
human activity, climate change and, 43
HVAC. *See* heating, ventilation, and air
    conditioning controls
hydrological systems, 43

Illuminating Engineering Society of
    North America (IESNA), 57
independent system operator (ISO), 72,
    109
information technology (IT), 9
installation capacity, 37
integrated project delivery (IPD), 50,
    90–91, 99
intelligent grids, 15
Intergovernmental Panel on Climate
    Change (IPCC), 43
internal combustion engines, 139
Internal Rates of Return (IRRs), 79, 83
inverters, 103, 133
investment tax credits, 133–34
IPCC. *See* Intergovernmental Panel on
    Climate Change
IPD. *See* integrated project delivery

IRRs. *See* Internal Rates of Return
Irvine Unified School District, 112,
    114
ISO. *See* independent system operator
IT. *See* information technology

Jeffrey Trail Middle School, 111–15,
    134

Keeling, Charles, 137
Kentucky Department of Education,
    123–26
Kern High School District, 20, 36, 80
kilowatt (kW), 20, 33, 36–38
kilowatt hour (kWh), 19–20, 35–38
Knox, Wyck, 106, 107
kW. *See* kilowatt
kWh. *See* kilowatt hour

labor costs, 18
Lake Powell, 132
Land o' Lakes school, 103
leadership, 142
Leadership in Energy and
    Environmental Design (LEED), 11;
    building codes and, 21; certification
    in, 16, 93–94; elementary school and,
    25; Northland Pines High School
    certified in, 80, 99, 143; platinum
    projects, 64; in public schools, 51;
    rating system of, 59; Schools Rating
    System of, 53; version 4, 65; VOC
    in, 100; zero energy solar schools
    and, 141
learning environment, 24, 139
lease agreements, 80
LEED. *See* Leadership in Energy and
    Environmental Design
light sensors, 71
light transmission, 63
lithium-ion batteries, 135
Living Building Challenge, 12, 53, 93,
    141
location challenges, 85–86, 97
Lovins, Amory, 59
Lovins, L. Hunter, 59

market share, utility pricing and, 37–38
Matthiessen, Lisa Fay, 53, 60
mean power profile, 67
mechanical, electrical, and plumbing
    systems (MEP), 25
mechanical systems, 85; chiller
    problems of, 67–68, 100–101;
    in commercial buildings, 78;
    geothermal systems as, 135–36;
    lower-cost, 114; in school buildings,
    10; VAV, 13
median costs, of school buildings,
    53–54
megawatt (MW), 5
MEP. *See* mechanical, electrical, and
    plumbing systems
methane, 40
Minnesota, 83
module prices, 33–34
monitoring and verification (M&V), 93
Morris, Peter, 53, 60
M&V. *See* monitoring and verification
MW. *See* megawatt

National Renewable Energy Laboratory
    (NREL), 33
National Technology Award, ASHRAE,
    105
natural gas, 34, 39–41, 85, 135
NBI. *See* New Buildings Institute
Nemet, Greg, 133
net metering, 88; buy-back rate of, 84;
    Minnesota law on, 83; of on-site
    solar PV systems, 36; power storage
    and, 73; in Virginia, 107, 112; in
    Wisconsin, 84; of zero energy solar
    schools, 83
net present values (NPVs), 79, 81
net zero energy, xii
New Buildings Institute (NBI), 32, 120,
    141
Newcastle Elementary, 118–20, 139
Northland Pines High School, 51, 59,
    81; battery storage and, 104; cash
    flow of, *82*; daily power profile of,
    67; energy costs of, 101, 105; glass

selection for, 57–58; HVAC system
    of, 62; LEED certification and, 80,
    99, 143; on-site solar PV installation
    of, 66–67; project elements of, 107;
    solar power of, 67, 139; TPI in, *82*;
    as zero energy solar school, 97–98
Northland Pines Middle School, 97–98
not-for-profit school districts, 58
NPVs. *See* net present values
NREL. *See* National Renewable Energy
    Laboratory

obstructionists, 61–63
Office of Energy Efficiency and
    Renewable Energy, 66
on-peak rates, 104
on-site solar PV systems: aggregate
    installed capacity of, 79; annual
    installation capacity of, 37; cap-
    and-trade system and, 134; cash-
    flow neutral, 58; challenges of,
    34–35; cost-saving opportunities
    of, 27; costs of, 33, 58, 88, 136;
    electrical grid connection with,
    68–69; elements of, 65; at high
    schools, 4–5; lease agreements for,
    80; net metering of, 36; Northland
    Pines High School installation of,
    66–67; performance drop-off and,
    108; in school buildings, 44; school
    design using, 132–33; for school
    districts, ix, 29; self-ballasted, 75; for
    sustainability, 12; TPI and, 12–13;
    virtual net metering of, 36; zero
    energy from, 36–37
operating costs, of energy, 23–24
OPR. *See* owner's project requirements
optimal design, of sustainable schools,
    74–75
optimization challenge, 109
over-lighting space, 60
owner's project requirements (OPR), 92

parking canopy solar panels, 114
peak-load time, 112
peak-use times, 71

per capita energy, ix
performance, cost and, 14–15
performance drop-off, 108
photovoltaic systems (PV), x, 20
planning process, 89
Plano Elementary School, 124
portaging, 117–18
Powerpack battery, 74
power profiles, *67*
power purchase agreements (PPAs), 20,
    32, 80, 90; in California, 112; school
    districts using, 101
power purchases, 84
PPAs. *See* power purchase agreements
price, school projects negotiating, 12
pricing, 33–34, 37–38, 40–41, 71
private schools, 143–44
professional building community, 52
Proposition 39 program. *See* California
    Proposition 39 Zero Net Energy
    Schools Pilot Program
public schools, 9, 17–18, 51
public utilities, 68
PV. *See* photovoltaic systems

rating system, of LEED, 59
R&D. *See* research and development
real-time pricing, 38, 71
RECs. *See* renewable energy credits
referendums: planning process and, 89;
    school administrators proposals for,
    89; school districts amount from,
    59, 88; school projects threshold for,
    91–92, 142
regional costs, 27–28, 56
regulations, 34, 37, 101
renewable energy, x, 12, 20; avoided
    costs in, 72; costs declining for, 139;
    in electrical grid, 38–39; Google and,
    138
renewable energy credits (RECs), 125
request for proposal (RFP), xiii, 51
research and development (R&D), x
residential buildings, 45
restrictive regulations, 101

retrofitting, of Newcastle Elementary,
    119
RFP. *See* request for proposal
Richardsville Elementary School, 107,
    123–24, 126
Richie, Mike, 98, 100–102, 143
River Crest Elementary School, 93;
    energy performance of, 35–36; glass
    selection for, 57–58; Source EUI in,
    25–26
river trips, xiii; decisions on, 87;
    leadership for, 142; planning, 31, 49;
    point of no return of, 55; portaging
    during, 117–18; predetermined
    destinations of, 50; river evaluation
    for, 21; technology used in, 77;
    unpredictability of, 131; where to go
    for, 23
roof design, of school buildings, 75
rooftop units, 113
Roosevelt, Teddy, 131
Rush, Denny, 118–19
Ryles, Mark, 124

school administrators, 78–79, 89
school buildings: buyers fear of,
    13; in California, 73; challenges
    with, 13–14; construction costs
    of, 52; design process of, 28, 75,
    132; electric power price for, 71;
    energy storage in, 69–70, 73–74;
    energy use of, 21, 32–34; glass
    selection for, 57; grading criteria
    of, 17; heating requirements of,
    39–40; high-performing, 25, 31;
    intelligence of, 71–72; mechanical
    systems in, 10; median costs of,
    53–54; objectives and design of,
    28; on-site solar PV systems in,
    44; real-time pricing for, 38; roof
    design of, 75; solar PV capacity of,
    84; solar PV systems for, 20, 132;
    supply chain for, 10; technologies
    emerging for, 4; zero energy in, 45.
    See also commercial buildings

school districts: building costs, 138, 144; competitive-cost energy in, 40; competitive solicitation process of, 98; demand side and, 2; design-bid-build process of, 90–91; design optimization for, 136; design process in, 1, 9–10, 90–91, 136; energy efficiency for, 24; financial risks taken by, 102–3; green design for, 1; lower-cost path for, 144; new or remodel issues in, 15; not-for-profit, 58; on-site solar PV systems in, xi, 29; PPAs used by, 101; project conditions in, 102; referendum amount for, 59, 88; sustainable buildings in, 32, 141; zero energy criteria in, 143–44. *See also specific school districts*

School Planning & Management, 60, 106

school projects: budgets of, 100; commissioning agents in, 78–79; construction process of, 14; cost disparities in, 11; critical points of, 55–56; design details of, 51; fund-raising for, 49–50; milestones in, 87; planning challenges for, 9; price negotiated in, 12; referendum threshold of, 91–92, 142; take-out point in, 92

School Rating System, of LEED, 53

science, technology, engineering, and mathematics (STEM), 20

sea levels, 43

seasonality barriers, 83

secondary schools, 28

SEIA. *See* Solar Energy Industry Association

Seibert, Kenneth, 123–25

seller side, 3

Site EUI, 19

smart buildings, 39

smart grid, 45, 112

smart school model, 112

Smith River, 137

solar capacity, 58, 90

solar energy: California power production of, 69; cost-competitive, 16; costs, xi–xii; diffusion model in, 20; firms adopting, 5–6; ground mounted, 36; of Jeffrey Trail Middle School, 114–15; Kern High School parking lot, 36; location challenges of, 97; multiyear construction process for, 140–41; of Northland Pines High School, 67, 139; panels, 63; parking canopy panels for, 114; powered schools, xii; US imposing import tariffs on, 37, 133

Solar Energy Industry Association (SEIA), 5, 20, 37

solar PV systems: ballasted, 119–20; cash flow from, 81; costs of, *33*, 101, 123, 125; in Discovery Elementary School, 105; economies-of-scale cost of, 120; on electrical grid, 1; financing terms for, 103; import tariffs on, 133; Newcastle Elementary capacity of, 139; parts of, 125–26; performance of, 93; positive financial returns from, 63–64, 81; power purchases from, 84; Richardsville Elementary School costs of, 123; for school buildings, 20, 132; school buildings capacity of, 84; TPI and, 108–9; US jobs in, 81. *See also* on-site solar PV systems

Source EUI, 19, 25–26, 32–33

Spirit Lake Community School District, 66

split incentives, 24

square footage, of public schools, 17–18

STEM. *See* science, technology, engineering, and mathematics

storage systems. *See* battery storage

student performance, 62–63

SunEdison, 112

supply chain, 10, 12

supply side, 3

sustainability: certification, 53; climate change solutions for, 29; criteria for, 24–25; design approach for, 53–54, 59–60, 74–75, 89–90; Discovery Elementary School goals of, 106, 109–10; of Jeffrey Trail Middle School, 114–15; of Newcastle Elementary, 120; on-site solar PV energy for, 12; zero energy design for, 53–54. *See also* renewable energy
sustainable buildings, 16, 32, 56, 61–62, 141
sustainable schools: climate change and, 140; costs of, 15, 56; financial barriers to, 77–78; integrated project delivery in, 50; optimal design of, 74–75; regional costs of, 27–28; solar powered, xii
system costs, of HVAC controls, 57

take-out point, 92
Target Source EUI, 28
tariff structures, 37, 71, 133
tax credits, 134
technology: battery, 134; for commercial and school buildings, 4; education influenced by, 62–63; in glass selection, 63; in green design, 5; information, 9; river trips using, 77
Tesla Powerpacks, 74
thin film PV system, 125
third-party investment (TPI), 20, 88; advantages of, 79; cash flows from, 82–83, 133–34; investment tax credits and, 133; in Northland Pines High School, *82*; obstructionists to, 63; on-site solar and, 12–13; solar PV systems and, 108–9; types of, 80; in Wisconsin, 27; of zero energy solar schools, 41
The Tipping Point (Gladwell), 138
TPI. *See* third-party investment
triple-pane windows, 108

Tunneling Through the Cost Barrier, 59
2030 Challenge, 45

union labor, 90
United States (US): cap-and-trade system in, 41; heating sources in, 19–20; per capita energy use of, ix; solar capacity in, 58; solar import tariffs imposed by, 37, 133; solar industry jobs in, 81; solar systems import tariffs by, 133
unpredictability, of river trips, 131
US. *See* United States
US Green Building Council (USGBC), 11
utilities: pricing, 37–38; rate structures, 37; regulation, 34

value engineering, 52
variable air volume (VAV), 13
VDMO (architectural firm), 107
Virginia, 107, 112
volitile organic compound (VOC), 100
voluntary economic response, 44

wind power, 66, 68–69
Wisconsin, 18, 80; avoided costs in, 72; net metering in, 84; restrictive regulations in, 101; seasonality barriers in, 83; TPI in, 27
Wisconsin Focus on Energy Program, 18, 99, 102
Wood McKenzie Power & Renewables, 37
worst-case scenario, 91, 100–101

ZE. *See* zero energy
ZEB. *See* Zero Energy Building
zero energy (ZE): building codes and design for, 142; capable, 111; elementary schools, 5; elements of, 74; Jeffrey Trail Middle School, 114; net, xii; from on-site solar PV systems, 36–37; residential buildings

with, 45; Richardsville Elementary School budget for, 126; in school buildings, 45; school districts criteria for, 143–44; sustainable design, 53–54

Zero Energy Building (ZEB), 32, 53

zero energy solar schools, xii, 1–2; area advantages for, 85–86; barriers to, 15; climate change and, 140; Discovery Elementary School as, 52; electrical grid connection to, 28–29; elementary schools, 5; heating needs of, 40; Jeffrey Trail Middle School, 114; LEED and, 141; natural gas options and, 85; net metering of, 83; Northland Pines High School as, 97–98; predetermined destinations of, 50; principles of, 72–73; regional cost averages of, 56; Richardsville Elementary School budget for, 126; solar capacity in, 90; TPI of, 41

zero net energy (ZNE), 119–20

Zero Net Energy School Retrofit Training, 20

ZNE. *See* zero net energy

About the Author

Mark Hanson's professional passion is the pursuit of sustainable, affordable zero energy solar schools and other commercial buildings. This pursuit is of a continuously evolving goal as technologies emerge, our understanding of our world and our impact on it grows, and economic forces and our collective values evolve. Energy resources, impact, and climate change are core drivers for the pursuit. Mark Hanson, PhD LEED BD+C, is Director of Sustainable Services at Hoffman Planning, Design & Construction, Inc. His work focuses on integrated design and project delivery, daylighting and lighting design, coordination of energy modeling and commissioning, HVAC systems review, solar energy, and monitoring and verification. Prior to joining Hoffman, Mark served as Executive Director of the Energy Center of Wisconsin from 1994 to 2001. He is a graduate of Harvard College with a BA in Economics and earned his master's degree in Water Resources Management and doctorate in Environmental Studies from the University of Wisconsin–Madison.

Mark's work has been published in academic journals such as *Environmental Management* and the *Journal of the American Planning Association.* An accomplished speaker, Mark's list of presentations on sustainability include "On Making Green Buildings the Rule" at the 2002 GreenBuild international conference and "Delivering Green Schools at Less than Conventional Costs" at the 2008 California Green Schools Summit. He has participated in 18 LEED-certified projects encompassing more than two million square feet with many attaining platinum and gold ratings.